Second Edition

# INTRODUCTORY BIOLOGICAL STATISTICS

Raymond E. Hampton
*late of Central Michigan University*
John E. Havel
*Missouri State University*

Long Grove, Illinois

For information about this book, contact:
Waveland Press, Inc.
4180 IL Route 83, Suite 101
Long Grove, IL 60047-9580
(847) 634-0081
info@waveland.com
www.waveland.com



10-digit ISBN 1-57766-380-2
13-digit ISBN 978-1-57766-380-5



Printed in the United States of America

8 7 6 5 4 3

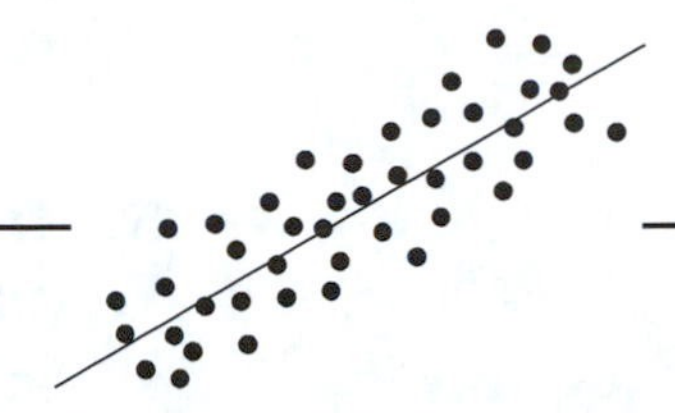

# Contents

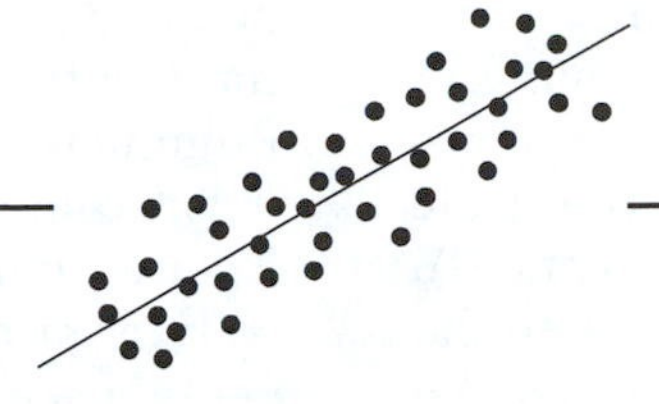

# Preface

The purpose of this book remains the same as for the first edition: to introduce upper-level undergraduates and beginning graduate students in biology to statistical analysis. We continue to maintain a pragmatic approach, illustrating the many uses of statistics in biology and other sciences, with a particular emphasis on a variety of inferential statistics. Statistics, like art and sport, takes practice to become skillful. We thus provide numerous examples and exercises. Some of the examples are made up, others are borrowed from colleagues or from our own work, and others are modified from other sources. Understanding the logic behind statistical procedures is very important, so we introduce students to the theory and state explicitly the assumptions underlying each method. We also show how these assumptions can be checked. We illustrate the calculation steps for most procedures and show computer output for most of the same examples. To maintain clarity of the text for introductory students, we have omitted some details, such as mathematical proofs. Students and instructors needing this information can consult more specialized texts.

This second edition has been thoroughly revised and updated. The introductions to each chapter have been rewritten to provide a clear rationale for why statistics is important. Some typographical errors have been corrected. To enable better connections between topics in different chapters, sections are now numbered. We also make occasional reference to other texts for further details on particular procedures, and provide a listing of these references in an appendix. To improve clarity of logically distinct topics, we have split several of the previous chapters. Data management and graphic presentations are now in separate chapters. The normal distribution is now treated separately from discrete probability distributions, and hypothesis testing is now in a separate chapter from estimation. Because of the length and complexity of analysis of variance, a thorough introduction to one-way ANOVA is presented in one chapter and its extension into other designs presented in a second chapter. Correlation is placed in a separate chapter after regression. We have added some new material. With descriptive statistics, we now introduce the important interquartile range and the display of skewed data in the form of Box plots. After introducing type I and type II errors, we provide a description of power analysis and use an example to demonstrate how the sample size can be determined for a minimum detectable difference. At the end of the book we briefly describe several new techniques (e.g., bootstrapping, meta-analysis), which students will encounter in the literature. To make room for the new material, we have deleted calculation steps for the two-way

ANOVA models and Friedman test and replaced these with worked examples done by computer. To enhance learning, graphics have been updated and new schematic diagrams added to many of the chapters. In our teaching of statistics, we have found it useful to summarize the logic behind selecting particular statistical tests. To help in this process, we also provide a "key to the tests" at the last chapter. We have updated the presentation of computer-generated statistics (using MINITAB version 14) and added useful graphics generated by this program (e.g., checks on normality and homogeneous variance assumptions). Finally, we have provided the larger data tables in digital format (as *Excel* and ASCII files on the accompanying CD), omitting some of the tedium of data entry.

## Acknowledgements

I am grateful to Chloe Hampton, for recommending that I revise Ray's book following his death, and to the numerous instructors who used the first edition and encouraged that the book be updated. I thank the expert reviewers for their comments on the book. Robert Angus, Mike Plummer, Jim Sumich, and Brenda Young provided comments on the first edition and Jamie Kneitel and Beatrice Snow reviewed the second edition prior to publication. My biometry students used a draft of the second edition and kindly pointed out errors. I also thank Don Rosso, my editor at Waveland Press, for his help on the many details of putting text and rough figures into publication. Finally, I thank my former professors and numerous colleagues for teaching me statistics through practice, my parents, Jerome and Doris Havel, for instilling in me an appreciation for numbers and clear writing, and my wife, Susan Robords, for helping me stay balanced.

This book is dedicated to the memory of Jerome Havel.

# 1

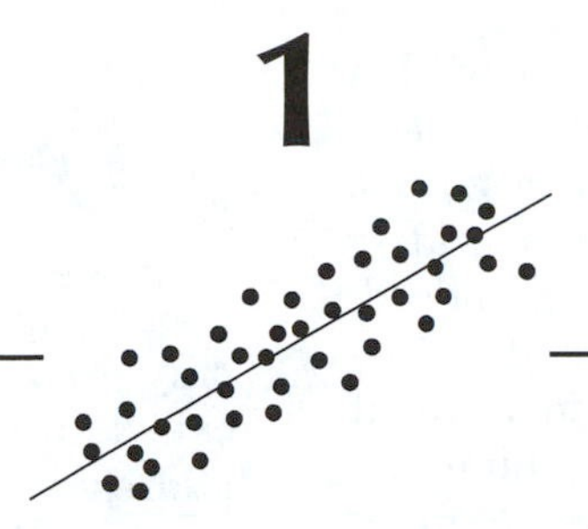

# Some Basic Concepts

Science is both an organized body of knowledge and a way of knowing. Any student of the sciences will agree that subjects like biology, chemistry, and physics contain a lot of information. What separates them from other kinds of knowledge is their process for generating new knowledge. All sciences progress by developing broad new generalizations (theories) that explain observable events of the physical universe. The essence of science is skepticism; any idea must be subject to evaluation with rational data. Theories are developed only after repeated tests of tentative explanations (hypotheses). By their nature, hypotheses must be subject to repeatable tests that can falsify them. In time, weak ideas are rejected and correct ideas are supported. Statistics plays an important role by helping us to more efficiently sort out the good ideas from the bad. Statistics also allows us to quantify our level of uncertainty about what we know.

## 1.1 What Is Statistics?

Statistics can refer simply to "a collection of numerical data." However, most scientists define **statistics** as "the mathematics of the collection, organization, and interpretation of numerical data and the analysis of population characteristics by inference from sampling" (*American Heritage Dictionary*). Statistics is the science of uncertainty, assigning probabilities to the reliability of estimates, to the reliability of conclusions, and to the likelihood of the outcome of future events. Statistics, at its best, is a way of thinking about much of the functioning of the physical universe.

For biologists, statistics is widely used in most sub-disciplines, from molecular biology to ecology. Pick up any professional journal and you are bound to run across some statistics. Statistics also is an important tool for practitioners of related disciplines. Statistics helps to track the causes of diseases in populations. Statistics is also widely used in product safety, such as guiding the efficient sampling of food for pathogenic microbes. Understanding statistics is also important for all educated citizens. Numerical data and conclusions about their meaning are often presented in the media, sometimes with a misleading message. A firm understanding of statistics provides readers with a healthy bit of skepticism, helping to sort weak and strong conclusions.

Statistics has several important uses to anyone else who investigates variable phenomena with probabilistic rather than deterministic outcomes, including biologists. First, it provides guidance for the organization and presentation of data. This is the realm of graphics and descriptive statistics, subjects covered in chapters 3 and

4. Another use of statistics is to determine, with a specified degree of uncertainty, if data mean what we think they do. This process refers to inferential statistics, which provides a widely accepted standard for testing hypotheses. Indeed, a key ingredient of the scientific method is this comparison of rational data with an idea about how we think nature works. We examine the nature of hypothesis testing in some detail in chapter 8 and explore a variety of methods in the remaining chapters. A final use of statistics is to provide a framework for experimental design, which guides the collection of future data and assures that we can reach some conclusion following a survey or experiment.

**Caution**

Many of us have spent time, energy, and money conducting experiments without giving much regard to the statistical analysis that will be applied to the collected data. Quite often, at the end, there is no test we can properly apply to our data because of the experimental design that was used (or because of the lack of one). Consequently, an inappropriate test is sometimes employed, either through ignorance or desperation. Biologists are thus well advised to think about statistical methods during the early phases of their study.

With any statistical analysis there are certain assumptions about the nature of the data and about how they were collected. It is not a bad idea to think of these assumptions as rules for when a particular test may or may not be properly used. If our experimental design and data do not follow these rules, we should not use the test! Usually, but not always, violations of these assumptions invalidate any conclusions based on the test. Thus, while it is possible, computationally, to apply many different tests to the same set of data, the validity of the conclusions based on any given test depends on how well the assumptions of the test were met, not on how well the test supports the conclusion we wish to support. Any research project employs certain methods of measuring things: we measure temperature with a thermometer, pH with a pH meter, and so on. Obviously, our conclusions are no better than the least accurate of our methods. Statistical inference is one of these methods. Applying an inappropriate statistical test to our data makes no more sense than measuring temperature with a pH meter!

## 1.2 Populations and Samples

Statistics is generally used to make inferences about populations based on data collected from only a portion of the population. That portion is called a **sample**. A question that is properly addressed by statistical methods is: how well does the sample represent the true situation of the entire population? In statistics the term **population** refers to all objects of a particular kind in the universe or in some designated subdivision of the universe. One must generally take care to specify the limits of the population being investigated.

**• Example 1.1**

A random sample from a population

The heights of a sample of 50 female students in a midwestern university were determined. The sample was selected from among all female students in the university by randomly picking their names from the student directory.

We might consider this measured group to be a sample of a larger population, but what population does it represent? Is the average height of this sample a reasonable estimate of the average height of all women everywhere? What about those groups of people who tend to be unusually tall or unusually short because of genetic and/or nutritional factors? Does our sample tell us anything about them? Hardly! The only population about which we can reasonably draw conclusions is the population of female students in this university, because this is the population from which our sample was drawn. Realistically, of course, if we are willing to assume that these college women are fairly typical in stature of midwestern women in their late teens to early twenties, we could generalize our conclusion to include this larger group. Note that the key phrase here is "willing to assume." Our generalization to a larger population than the one that was sampled would be accurate only if our assumption is correct! In a purely statistical sense, we have no basis for concluding anything about the height of women who are not members of the sampled population. Imagine that, unknown to us, taller women are more likely to attend a university. Our assumption that the population sampled is representative of some larger population would be incorrect, and our generalization to the larger population would therefore be inappropriate and misleading.

## 1.3 RANDOMNESS

Generally, when we wish to make inferences about a population from a sample of that population, it is important that the individuals selected to make up the sample be chosen at random. Randomness does not imply casual, haphazard, or unplanned. Rather, **randomness** means that each possible sample of the same size that could conceivably be drawn from this population has an equal probability of being drawn. When this is not the case—that is, when a sample is not randomly selected—the sample is said to be biased. In example 1.1, if our intention had been to conclude something about the average height of *all* women, our sample would be likely have been biased because it included individuals (observations) from only one specific group of women; therefore, not all possible samples of this size from the population of all women everywhere could have been drawn. Any inference that we might make about the population based on our sample is of no value.

How does one obtain a random sample? There are a number of ways to do this. All of us have drawn names or numbers from a hat, or we have seen a commercial for a lottery where one numbered ball is selected from a drum. The process is designed so that each number or name has the same chance of being drawn. Computers are now widely used to perform similar procedures. Another method, which helps illustrate the process, is the use of a random number table (*see* appendix A, table A.10). In example 1.1 we could have assigned a number to each woman in the university and then consulted the random number table to decide which individuals to include in the sample. We might do this by closing our eyes and touching the table with a pencil to select a starting number and then using consecutive numbers from left to right, right to left, top to bottom, or any other sequence. The fact that each group of numbers in this table consists of 5 digits is of no significance. This simply makes the table a bit easier to read. If we wish to select random numbers of three digits each, we simply read them in groups of three.

• **Example 1.2**

A biased sample from a population

A large university wanted to illustrate the success of its graduates by publicizing the average annual income of its former students. They selected alumni 10 years after graduation and attempted to contact each individual by telephone. Although able to contact only 20% of this group, the university averaged the results and reported that their alumni earned over twice the average for this age group in the general population. (The university thus assumed or implied that this sample of 20% was a random sample from the entire population.)

Telephone surveys are widely used in the social sciences and, at first glance, seem a good way to obtain the alumni income data. In this case, the university sampled 20% of this alumni group for their incomes. However, this was not a random sample, as the 80% missing from the sample constituted those alumni who could not be reached (or would not answer the question). Those responding to the survey were likely a biased sample from the population of interest. We could imagine that people who were easier to reach by phone would include a higher fraction of professionals and those listed in who's who directories. The survey would be less likely to reach those who were homeless or substance abusers, groups which are less likely to earn larger incomes.

Biased sampling is also a problem in biology. For example, snake populations are sometimes estimated by conducting road surveys. The number of snakes spotted crossing the road should be proportional to population size. However, biologists must be careful in how they interpret such data, since this survey technique is biased by activity levels of the snakes. For instance, in the spring nearly all snakes observed crossing the road are adult males. We might be tempted to conclude that this population has very few females, when in fact we are counting only active individuals. In spring, only the adult males are actively searching for mates!

Consider one last example. The hellbender salamander, once common in Ozarks streams, is now locally threatened and of considerable conservation interest. Body-size distributions from recent collections show very few individuals in small age classes, suggesting either a failure of adults to reproduce or high death rates of juveniles. However, collection methods tend to focus on higher depths and small juveniles may burrow deeply in the gravel. Thus collections may be biased against collecting small individuals. Nevertheless, it is quite reasonable to compare population estimates and age distributions over time when the same sampling techniques have been used.

An important rule for the interpretation of statistics is to be aware of the population of interest versus the population actually sampled. When these fail to agree, we either need to change our sampling strategy or modify our conclusions about the results. For the snake and hellbender examples above, making this distinction about sampling requires knowledge about the natural history of the animal.

Try consulting the primary research literature in your particular area of interest to see how scientists have obtained random samples. What is the statistical population? How was randomness assured?

## 1.4 Independence

In addition to the requirement that members of a sample be randomly chosen, most statistical procedures assume that each sampled unit is independent of others. **Independence** means that the choice of any one individual for inclusion in a sample does not in any way change the probability that any other individual will be chosen, or that the occurrence of some event in our sample does not in any way influence the outcome of subsequent events in our sample.

In example 1.1, 50 female students were selected at random from the student directory. Imagine that, to insure randomness, each woman's name in the entire population of university women was written on a slip of paper and placed in a box. Fifty names were then drawn from the box, which received a thorough shaking between each draw. This is a cumbersome but effective way of achieving randomness, and we may feel confident that we have selected a random sample from the population. But are the observations independent? If each name that was drawn from the box was returned to the box before the next draw (a process known as sampling with replacement), the answer is yes. On the other hand, if sampling had been without replacement (if the names drawn were not returned to the box), then the observations would not be independent. Why is this? Imagine that the entire population of women in this university consists of 1,000 individuals. On the first draw, any individual has a probability of 1/1000 (0.001000) of being selected. However, on the second draw, there are now only 999 individuals remaining in the population who may be selected, and the probability that any individual will be selected is 1/999 (0.001001). By the time the fiftieth individual is selected, the probability that any individual will be selected is 1/951 (0.001052). Thus, each time an individual is removed from a population as one of a sample and is not replaced, the probability that any of the remaining individuals will be selected increases, and the sample technically is not independent. Note, however, that the change in this probability is very small in the case outlined above, and no real harm is done to the requirement of independence. Generally, when sample size is small compared to population size, one need not be concerned about the lack of independence that results from sampling without replacement. It is a problem, however, when the sample size is fairly large compared to the overall population size. In such a situation, sampling with replacement might be used, or one might wish to apply what is known as the finite population correction. (However, since this is an uncommon situation, the details of the finite population correction are not given here.) A frequent violation of the assumption of independence is when a researcher makes repeated measurements using the same experimental creature (or event) or makes repeated measurements on each of several creatures (or events).

---

- **Example 1.3**

  A sample in which observations are not independent

An ornithologist wished to know if goldfinches prefer sunflower seeds or thistle seeds. Accordingly, two feeders were set up, one with sunflower seeds and one with thistle seeds. She observed that one goldfinch visited the sunflower-seed feeder 50 times and the thistle-seed feeder 10 times in a certain period of time.

---

From these observations we might be tempted to conclude that goldfinches therefore prefer sunflower seeds, and we could even conduct a statistical test (e.g., a chi-square test) that would support our conclusion. To conduct this test, we would assume a sample size of 60, since there were 60 total visits to the feeders. In fact, the sample size is one, since we observed only one bird, and our statistical test would be meaningless! The only thing that we could logically conclude is that this one goldfinch preferred sunflower seeds; we could conclude nothing about the entire population of goldfinches.

Consider a slightly more complicated version of this same situation. We tabulate visits to the feeder as before, but this time there are several birds involved. However, we do not record which bird we are counting at any particular time. It is possible—even very likely—that each bird was counted several times and that some birds were counted more often than others. Nevertheless, we then pool all of our observations to make a large sample. In this case, the sample size is not the total number of visits we recorded but the number of birds involved. Since we don't know how many different birds were involved, we don't know the sample size. This is sometimes called the "pooling fallacy," and it is almost always inappropriate. There are several correct ways of conducting this experiment depending on the statistical test we wish to use, but the two scenarios outlined above are both incorrect because they violate the assumption of independence.

## 1.5 Other Types of Samples

The type of sample that we have been considering is called a simple random sample, and this is perhaps the most common type of sample used in statistical inference. It is not, however, the only type of sample that may be used. For instance, we occasionally divide an environment or individuals into groups first and then sample from each group. Such sampling occurs when a blood pressure medication is tested simultaneously on different age groups. Sampling can be a complex subject, so researchers typically carefully read the literature in their discipline and ask statistical consultants for advice when starting a new study.

**Caution**

A very common mistake that one encounters in the use of statistical tests, at least in the biological literature, is a lack of independence of observations. "Real life" scenarios like our goldfinch example above are surprisingly common. The violation of independence of observations, when this is an assumption of the statistical test being used, is a very serious error.

## Key Terms

independence
population
randomness
sample
statistics

## Exercises

1.1 Using table A.10 (appendix A) to generate random numbers, select three simple random samples of 10, 20, and 30 bluegill sunfish from the data in digital appendix 1. For purposes of this exercise, we consider the data in this table to be the entire population of interest. To take your samples, select an arbitrary starting place in the random number table, such as by closing your eyes and touching the table with a pencil. Next, record sequences of three digits as they occur, skipping any three-digit sequences that are larger than the number of observations in the Bluegill population. When you have selected 10, 20, or 30 such random numbers, record the lengths of individual fish specified by these numbers. Use a different starting place in the random number table for each sample, and avoid the temptation of obtaining your sample of 30 by combining your samples of 10 and 20! Save the data so obtained for future use. Each student in the class should obtain different samples for reasons that will be apparent later.

1.2 If you have statistical software available, use the computer to gather similar random samples from the bluegill population. (Your instructor may supply instructions for this procedure, depending on the software package.)

1.3 Using the procedures in exercise 1.1 and/or 1.2, select simple random samples of 10, 20, and 30 male mosquito fish lengths from the data in digital appendix table 2. Collect similar random samples of female mosquito fish lengths. Notice that the data show a code for sex.

1.4 Using the procedures in exercise 1.1 and/or 1.2, select simple random samples of 10, 20, and 30 resting pulse rates of general biology students from the data in digital appendix table 3.

1.5 Using the procedures in exercise 1.1 and/or 1.2, select simple random samples of 10, 20, and 30 reaction times of general biology students from the data in digital appendix table 3.

# 2

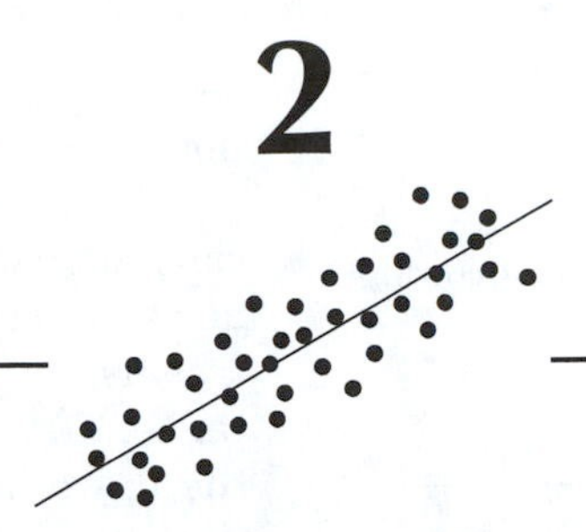

# Data Measurement and Management of Numbers

Before we get too far with statistics, it's important that we develop a common vocabulary about the sorts of things we measure and count, as well as about how the numbers that are generated can be organized. In this chapter we examine the different sorts of variables encountered in statistics, how they are represented, and their scales of measurement. We also briefly show how data can be managed in an efficient manner. In chapters 3 and 4 we examine different ways numbers are summarized and presented through graphics and descriptive statistics. A solid foundation of data description makes hypothesis testing much more understandable, both in this course and as a general practice.

## 2.1 Variables and Data

**Variables** are characteristics that may differ from one member of a population to the next. **Data** are the values of these variables for individual members of the population. Variables can be grouped according to several characteristics.

### 2.1.1 Types of Variables

Variables can be grouped according to several characteristics, such as whether they represent measurements, ranks, or simple categories. Measurements may be either discrete or continuous. Consider first a couple of examples.

---

• **Example 2.1**
A continuous variable

The heights of five pine trees in a certain area of forest were measured, with these results:

| Tree Number | 1 | 2 | 3 | 4 | 5 |
|---|---|---|---|---|---|
| Height (m) | 31.3 | 29.1 | 32.6 | 19.5 | 37.8 |

---

In these observations tree height is a variable. A variable such as this is a **continuous variable** because it may assume any imaginable value within a certain range. Measurements of length, mass, time, temperature, concentration, and so on are examples of continuous variables.

In other situations, we may collect data on phenomena that occur in only discrete steps or units, where intermediate values are not possible, as in example 2.2.

• **Example 2.2**

A discrete variable

In investigating the litter size of garter snakes, the following data were collected:

| Snake Number | 1 | 2 | 3 | 4 | 5 |
|---|---|---|---|---|---|
| Litter Size | 5 | 8 | 6 | 9 | 7 |

Since fractions of baby snakes do not exist, we would never expect to obtain a measurement that is not a whole number. Variables of this kind are called **discrete variables**.

Still other variables cannot be measured on a scale in which the intervals or units have a consistent relationship to each other, yet the observations made on such variables can be ranked with respect to their relative magnitude. These are called **ranked variables**. The position of an individual in a dominance hierarchy, in which individuals may be ranked with respect to their relative position in the group, is a good example of a ranked variable. If individual A is dominant over individual B, who is in turn dominant over individual C, we may rank the individuals as A>B>C, or we may assign a number to each that indicates their relative position. Thus, A would have a rank of 3, B a rank of 2, and C a rank of 1. Note, however, that the difference between 3 and 2 need not be equal to the difference between 2 and 1. The only relationships that are expressed in ranked variables are "equal to," "greater than," or "less than."

Variables that can neither be expressed quantitatively nor ranked with respect to relative magnitude are called **attributes**. Male or female, red or white, alive or dead, and so on, are examples of attributes. One may designate the class of objects to which an individual belongs, but values such as "greater than" or "less than" have no validity when applied to such attributes.

Two other variable types are occasionally encountered. One is called a **transformed variable**. Such a variable is the result of performing some mathematical operation on an original variable or of having a measuring device perform such an operation. An example is the measurement of pH, which is the negative logarithm of the hydrogen-ion concentration in a solution ($pH = -\log[H^+]$). The actual variable being measured is the hydrogen-ion concentration. The transformed variable is pH. Although we don't measure these types of variables very often, mathematical transformations are frequently used during advanced statistical analyses (e.g., see chapter 10: ANOVA, when assumptions are not met).

**Derived variables** are computed from two continuous measurements. Many rates are derived variables. If we measure how far a snail travels, we have measured a variable called distance. If we measure how long it takes to get there, we have measured another variable called time. If we combine these two variables and measure how far the snail travels divided by the time of its trip, we have now measured a derived variable called velocity. Examples of derived variables are ratios, percentages, and rates.

Finally, variables are sometimes described in terms of how we think they interact with other variables (the stuff of hypothesis testing!). **Response variables** are those whose variation we suspect depends on some other variable, called a **predictor variable**. For instance, in regression analysis (chapter 12) we may be interested in testing the influence of drug dosage on blood pressure. Here, blood pressure is the response variable and drug dosage the predictor. You may have also referred to these as dependent and independent variables, respectively. Both sets of terms are correct.

Notice that we have described variables in terms of their fundamental types and by their relationships. In section 2.2 below we also examine different scales of measurement. All these descriptions of variables and scales are part of the common vocabulary of statistics, so you will see these terms frequently throughout this book.

### 2.1.2 Significant Figures and Rounding Rules

Before we go further, let's consider a common practical problem with reporting the values of continuous measurements. Occasionally, following some calculations we are left with a large number of digits. For instance, suppose we are interested in the respiratory physiology of elephants and have made the following measurements. Big Burt, who tips the scales at 1043 kg, consumes 202 liters of oxygen per minute (202 L $O_2$ min.$^{-1}$). According to my calculator, his weight-specific respiratory rate is 0.1936721 L $O_2$ min.$^{-1}$ kg$^{-1}$. (Notice that this is an example of a derived variable, which used two different continuous measurements, weight and respiratory rate, for its calculation.) Despite what my calculator reports, the weight-specific respiratory rate should be reported only to three significant figures. Whatever measurement has the fewest significant figures sets the limit to the number of significant figures in the derived variable. In the case of Burt, his weight-specific respiratory rate should be reported as 0.194 L $O_2$ $min.^{-1}$ $kg^{-1}$. Notice that the leading zero doesn't count as a significant figure, since this number could just have well been written in scientific notation as $1.94 \times 10^{-1}$.

Rounding numbers follows some simple rules, which are best demonstrated with examples. Notice the conditions where we "round up" (add one to the last significant figure) and "round down" (keep the last significant figure the same).

**Table 2.1 Demonstration of rounding rules**

| Original number | Rounded to 3 signif. figures | Rounded to 2 signif. figures |
|---|---|---|
| 1.852 | 1.85 | 1.9 |
| 1.856 | 1.86 | 1.9 |
| 1.8551 | 1.86 | 1.9 |
| 1.855 | 1.86 | 1.9 |
| 1.845 | 1.84 | 1.8 |
| 1.865 | 1.86 | 1.9 |
| 0.1852 | 0.185 | 0.19 |

When performing a series of calculations, *save all the numbers in memory* (in the calculator or computer) until the final answer is complete. Common practice is to first show the answer with more than enough digits and then round to the proper number of significant figures. As you do practice exercises from this book, you will notice that the answer key often shows more than the correct number of significant figures. This makes it easier for you to know that your answers are on the right track. In a formal report, we would round the final answers.

### 2.1.3 Data and Their Representation

**Data** are individual measurements or observations of a variable made on individual units of the population under study. Data are also sometimes called variants or observations. In the pine tree example, each tree is a unit of the population; height is the variable; and the height of an individual tree is a datum, variant, or observation. A variable is conventionally designated with a letter, typically $x$, and individual data (observations, variants) as $x_1, x_2, x_3, \ldots, x_n$, where $x_1$ is the first datum, $x_2$ the second, and $x_n$ the $n$th observation in the series. When two variables are under study, one is usually designated as $x$ and the other as $y$. But this convention is not sacrosanct! Professionals working with numbers might represent variables by other letters; however, a variable must remain consistent once it has been defined. The total number of observations of a variable in any given sample, the **sample size**, is designated as $\boldsymbol{n}$.

## 2.2 Scales of Measurement

Most objects or events have at least one characteristic that is detectable and therefore measurable. Four levels, or scales, of measurement are recognized: the nominal scale, the ordinal scale, the interval scale, and the ratio scale. Each has certain distinguishing characteristics, and each may be manipulated only in certain ways.

### 2.2.1 The Nominal Scale

The **nominal scale**, sometimes called the classificatory scale, uses numbers or other symbols to classify objects or events into one of two or more mutually exclusive categories. For example, each individual in a population might be "measured" to be either a male or a female. No individual may assume an intermediate value between any two categories, nor may it be a member of more than one category. The only relationships between individuals, with respect to a variable measured on a nominal scale, are "equal" and "not equal." "Greater than" and "less than" have no meaning in the nominal scale of measurement. Attributes such as male or female, present or absent, and so on are measured on a nominal scale.

### 2.2.2 The Ordinal (Ranking) Scale

In some instances, it is possible to distinguish objects in different classes and determine whether objects in one class are either "greater than" or "less than" objects in another class. When it is possible to determine that an object from group A is in some respect greater than or less than some object from group B, but it is not possible to determine how much the two objects differ, we have measured on an **ordinal** scale. An example might help. Suppose that a wildlife biologist is studying the dominance hierarchy in male lions. (Their dominance is important for such things as controlling territories and access to mates.) The biologist records contests she observes between pairs of males and scores who wins each contest. Based on these data, she places the males in this order, from most dominant to most submissive: Yellowhead > Spooky > Bigfoot > Shyone. In such ordinal measurements, the actual distance between any two adjacent categories (in this case, names of lions) is unknown and need not be the same between all adjacent pairs. For instance, Yellowhead may be just barely dominant over Spooky, while Spooky is strongly dominant over Bigfoot. Customarily, such measurements are "scored" using numerical symbols. The "smallest" or lowest ranking item ("Shyone," in this example) receives a score of 1, the next smallest ("Bigfoot") a score of 2, and so on. "Yellowhead," in our example, would receive a score of 4. Ranked variables are always measured on an ordinal scale.

### 2.2.3 The Interval Scale and the Ratio Scale

The **interval scale** and the **ratio scale** both have a constant and defined unit of measurement, which assigns a real number to all measurements made on these scales. Using either of these scales, not only may we specify that one object is either greater than or less than some other object, but we may specify how much greater or less. The relationship between numbers on these scales is consistent. In other words, the interval between 1 and 2 is the same as the difference between 3 and 4 or between any other pair of adjacent numbers. The difference between these scales is that an interval scale uses an arbitrary zero point, while a ratio scale has a true zero point. An example of the use of an interval scale is the measurement of temperature using the Celsius scale. In this scale the difference in temperature of freezing water and boiling water is arbitrarily divided into 100 equal units, each of which is called a degree, and the zero point is arbitrarily chosen as the freezing point of water. The ratio scale, on the other hand, has a true zero point, and in this case zero means there is none of the measured attribute present. Measurements of mass, length, time, velocity, and so on, utilize a ratio scale. All of the statistical tests that we consider appropriate for use with interval measurements are also appropriate for use with ratio measurements and vice versa. However, when ratio or interval measurement is one of the assumptions of a particular test, it may not be used with ordinal or nominal measurements. Continuous and discrete measurement variables are measured on an interval or ratio scale.

In summary, we have identified basic variable types and the scales with which they are measured. Examples are shown below (table 2.2). You should be able to think of other examples.

**Table 2.2 Some Examples of Variable Types and Their Scales of Measurement**

| Variable Type | Scale of Measurement | Examples |
|---|---|---|
| Attribute | Nominal | Fruit fly phenotype: white eye or wild type |
| | | Sex of snake: male or female |
| Ranks | Ordinal | Status of patient: weaker, stable, better |
| | | Pollination sequence: first, second, third |
| Discrete Measurement | Ratio | Number of points on deer antlers |
| | | Number of fleas on a dog |
| Continuous Measurement | Interval | Body temperature (°C) |
| | Ratio | Weight of a warthog (kg) |

**Caution**

Many of the most commonly used statistical tests are based on assuming (among other things) measurement on an interval or ratio scale. For measurement on an ordinal or nominal scale, the use of tests that assume interval or ratio measurement is always inappropriate.

### 2.2.4 Converting Data from One Scale to Another

It is not possible to convert measurements made on a nominal scale to ordinal, interval, or ratio scales, nor is it possible to convert ordinal measurements to interval or ratio scales. However, it is possible to go the other way. Data measured on interval or ratio scales may be converted to ordinal or nominal scales, and data measured on an ordinal scale may be converted to a nominal scale. Consider example 2.1, which is reproduced here:

| Tree Number | 1 | 2 | 3 | 4 | 5 |
|---|---|---|---|---|---|
| Height (m) | 31.3 | 29.1 | 32.6 | 19.5 | 37.8 |

Tree height is a continuous variable measured on a ratio scale. Suppose we wanted to convert these measurements to an ordinal scale. The shortest tree in the group is tree 4, so we assign it a rank of 1. The next shortest is tree 2, which we give a rank of 2, and so on, until we have ranked each tree with respect to the others. Tree 5, being the tallest, is given +a rank of 5. Our data now look like this:

| Tree Number | 1 | 2 | 3 | 4 | 5 |
|---|---|---|---|---|---|
| Height rank | 3 | 2 | 4 | 1 | 5 |

Note that when we do this sort of conversion, we lose some information originally contained in our data. We now know only which tree is taller or shorter than which others. We do not know how much shorter or taller, nor do we know the height of any individual tree.

In subsequent chapters we discuss a number of situations in which such conversions of data from ratio scale to ordinal scale is appropriate. In particular, nonparametric statistics (e.g., chapter 9, Mann-Whitney test) are calculated after converting measurements to ranks.

## 2.3 Data Management

While studying statistics, you will have access to calculators, spreadsheets, and software packages dedicated to statistics. These modern inventions are a step up from the abacus and slide rule, but they still don't think very well! That's your job. Setting up the problem on paper or in your head is always the critical first step. Remember that computers are very efficient at giving wrong answers as well as right answers. Being able to quickly spot errors is a very important skill for students (and their instructors). One way to do that is to make a "ballpark estimate" ahead of time and then see whether later answers make sense. A simple graph is very helpful in this regard. In chapter 3, we see how to construct a histogram, which allows us to visualize the average and variation in a set of data.

When working with numbers, one has a choice among many different tools. Sometimes hand calculations are simplest, and these may be the only method available, such as while thinking in a restaurant or trying to impress your boss! A cheap scientific calculator is useful for doing operations on small masses of data and will be indispensable for working problems in this book. Besides the usual arithmetic operations, knowing how to use storage memories and parentheses can save you much time and reduce the chance of rounding errors. Although these calculators also have some statistical functions, computers are much better. Spreadsheets are very useful for organizing numbers, preparing tables, and performing simple repeated operations. These packages also create graphs and perform inferential statistical analysis, but not very well. Numerous computer packages dedicated to statistics are now available (e.g., SAS, SPSS). For this book, we have used *Minitab Statistical Software* (version 14), which is widely available, versatile, and easy to use. Such statistical packages do data manipulations, descriptive statistics, and a wide variety of inferential tests. Statistical packages also graph sufficiently for the purposes of this book. Most professionals use a number of different programs. For instance, within a *Windows* landscape, one might manage numbers and create tables in *Excel*, copy and paste numbers into *Minitab* to run statistical analyses, and then copy and paste tables and statistical output into *Word* to support writing a report. (By the way, we did these things repeatedly in preparing this book!) Copying and pasting numbers into a dedicated graphics package, such as *Sigmaplot*, allows preparation of publication-quality graphics.

### Key Terms

| | |
|---|---|
| attribute | ranked variable |
| continuous variable | ratio scale |
| data (datum) | response variable |
| derived variable | sample size ($n$) |
| discrete variable | significant figure |
| interval scale | transformed variable |
| nominal scale | variable |
| ordinal scale | variant |
| predictor variable | |

### Exercises

2.1 Following are the heights (in centimeters) of a small sample of male humans:

187 171 181 180 178 171 174 177 172 178 182

a) What is the scale of measurement by which this variable was measured?

b) Is this a discrete or continuous variable?

2.2 Following are the lengths (in millimeters) of a small sample of largemouth bass:

210 325 285 402 350 240 409 330 295 325

a) What is the scale of measurement by which this variable was measured?

b) Is this a continuous or discrete variable?

2.3 An ecology class determined the number of ant-lion pits in a sample of 100 randomly selected one-meter-square quadrats. The results were as follows:

| Pits/Quadrat ($x$) | Number of Quadrats Containing $x$ Pits (Frequency) |
|---|---|
| 0 | 5 |
| 1 | 15 |
| 2 | 23 |
| 3 | 21 |
| 4 | 17 |
| 5 | 11 |
| 6 | 5 |
| 7 | 2 |
| 8 | 1 |

a) What is the scale of measurement by which this variable ($x$) was measured?

b) Is this variable discrete or continuous?

2.4 Garter snakes respond to an overhead moving object by exhibiting an "escape" response. The intensity of this response may be measured on a somewhat subjective scale ranging from 3 to 0. A rapid movement of the snake from one location to another is given a score of 3, a movement involving at least one-third of the snake's body but not resulting in relocation of the animal is given a score of 2, a slight movement of the head only is given a score of 1, and no visible response is given a score of 0. Fifteen snakes were tested for this response and received the following scores. What scale of measurement was used to obtain these data?

| | | |
|---|---|---|
| 3 | 2 | 3 |
| 1 | 3 | 2 |
| 3 | 3 | 1 |
| 0 | 2 | 3 |
| 2 | 0 | 2 |

2.5 Convert the following male heights (given in centimeters) into an ordinal scale.

| Male Number | 1 | 2 | 3 | 4 | 5 | 6 |
|---|---|---|---|---|---|---|
| Height | 181 | 202 | 190 | 185 | 190 | 200 |

2.6 Classify any male from exercise 2.5 who is less than 200 cm to be "short" and any male who is 200 cm or taller to be "tall," and convert the data into a nominal measurement.

2.7 Convert the following number of bullfrogs per pond into an ordinal scale.

| Pond Number | 1 | 2 | 3 | 4 | 5 | 6 | 7 | 8 | 9 | 10 | 11 |
|---|---|---|---|---|---|---|---|---|---|---|---|
| Bullfrogs | 34 | 65 | 23 | 34 | 18 | 20 | 15 | 70 | 15 | 18 | 34 |

2.8 Classify any pond from exercise 2.7 that has 30 or fewer bullfrogs as "sparsely populated" and any pond that has more than 30 bullfrogs as "densely populated," and convert the data into a nominal scale.

2.9 For the following situations, identify the type of variable, indicate the scale of measurement, and indicate those variables that are derived or transformed.

a) The velocity of an enzyme-catalyzed reaction of substrate in micromoles converted per milligram of protein per minute.

b) The number of male bullfrogs in a pond.

c) Sex of a salamander: male, female, intersex.

d) pH and $pK_a$.

e) The velocity of snails, in furlongs per fortnight.

2.10 Using conventional rounding rules, round the following numbers to three significant figures. (Hint: it is sometimes helpful to express in scientific notation.)

a) 106.55

b) 0.06819

c) 3.0495

d) 7815.01

e) 2.9149

f) 20.1500

2.11 Choose a research article from a field of your interest that includes numerical data. What types of variables were studied? What scales of measurement were used? If measurements were on the ratio (or interval) scale, are these discrete or continuous measurements? If they report the information, what computer software did they use to analyze their data?

# 3

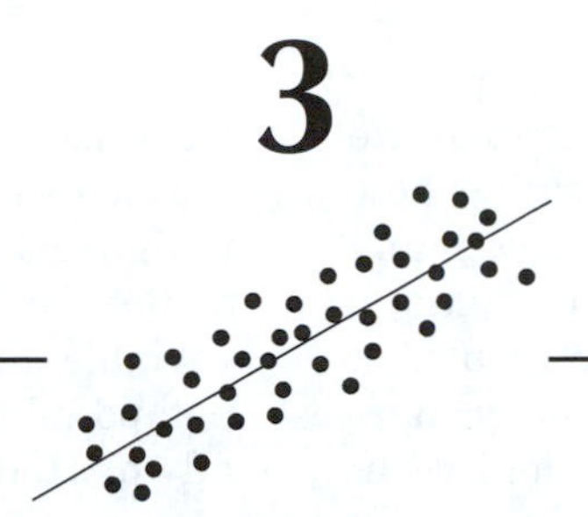

# Frequency Distributions and Graphic Presentation of Data

Large quantities of data may be difficult to interpret. In this chapter and the next we consider some basic techniques for organizing and presenting data to make them more readily understandable. Graphic techniques are widely used by investigators for data exploration and also to display summary data in published literature. We focus on frequency distributions in this chapter and then briefly introduce some other graph types that appear in later chapters.

## 3.1 Frequency Distributions of Discrete Variables

**Frequency distributions** display a summary of the number of observations that take on specific values or ranges of values. The **frequency** is a count of the number of observations in a category. Frequency distributions may be shown either as a table or a graph, and the data summarized can be from either discrete or continuous variables.

You will recall that discrete variables may assume only certain values and that intermediate values are not possible. The following example presents raw data for a discrete variable and then shows how these data may be organized.

### • Example 3.1

Frequency distribution of a discrete variable

The raw data in table 3.1 show the number of maple seedlings that were present in 100 one-meter-square quadrats. (A quadrat is a square sampling unit used by ecologists to measure the abundance and distribution of plants and stationary animals.)

**Table 3.1** Maple seedlings per one-meter-square quadrat

| | | | | | | | | | |
|---|---|---|---|---|---|---|---|---|---|
| 0 | 1 | 0 | 2 | 0 | 0 | 1 | 2 | 1 | 1 |
| 0 | 1 | 2 | 5 | 1 | 0 | 2 | 1 | 1 | 2 |
| 1 | 2 | 3 | 4 | 3 | 4 | 2 | 1 | 0 | 5 |
| 0 | 0 | 2 | 0 | 1 | 0 | 0 | 0 | 2 | 0 |
| 1 | 3 | 4 | 2 | 1 | 0 | 0 | 1 | 0 | 1 |
| 2 | 3 | 5 | 0 | 0 | 0 | 1 | 2 | 3 | 4 |
| 3 | 3 | 0 | 2 | 0 | 4 | 5 | 0 | 0 | 0 |
| 0 | 1 | 0 | 0 | 1 | 0 | 1 | 0 | 0 | 1 |
| 0 | 1 | 0 | 1 | 3 | 1 | 5 | 1 | 1 | 3 |
| 4 | 3 | 2 | 0 | 0 | 4 | 1 | 2 | 1 | 1 |

Data from M. Hamas

Clearly data in this form are rather difficult to interpret. From glancing through the numbers it would seem that the quadrats contained about 2 or 3 plants each, except that many contained no plants at all. Grouping the data into a frequency distribution can serve as a useful first step in the analysis. In this case we would determine how many quadrats contained 0 plants, how many contained 1 plant, how many contained 2 plants, and so on. We would then tabulate the results in the manner shown in table 3.2. This arrangement makes the data much easier to deal with. We can see at a glance that 35 quadrats contained 0 seedlings, 28 contained only 1, and so on. A frequency distribution is often represented in the form of a bar graph, which makes the information even more immediately accessible. Figure 3.1 is such a bar graph, constructed from the frequency distribution in table 3.2.

**Table 3.2** Maple seedlings per quadrat

| Number of Plants/Quadrat | Frequency |
|---|---|
| 0 | 35 |
| 1 | 28 |
| 2 | 15 |
| 3 | 10 |
| 4 | 7 |
| 5 | 5 |

In effect we are plotting a variable along the $x$-axis (in this case, the number of plants per quadrat) against the number of times that variable occurred (its frequency) on the $y$-axis. Graphs of frequency distributions are often called **histograms** if the variable graphed is continuous and **bar graphs** if the variable is discontinuous (discrete, as in this case). We will have many occasions to make use of them. Note that the bars in figure 3.1 representing the various frequencies do not touch each other. This is because a discrete variable is involved, and intermediate values are not possible. Discrete variables are conventionally plotted this way.

**Figure 3.1** Maple seedlings per 1-square-meter quadrat (an example of a bar graph)

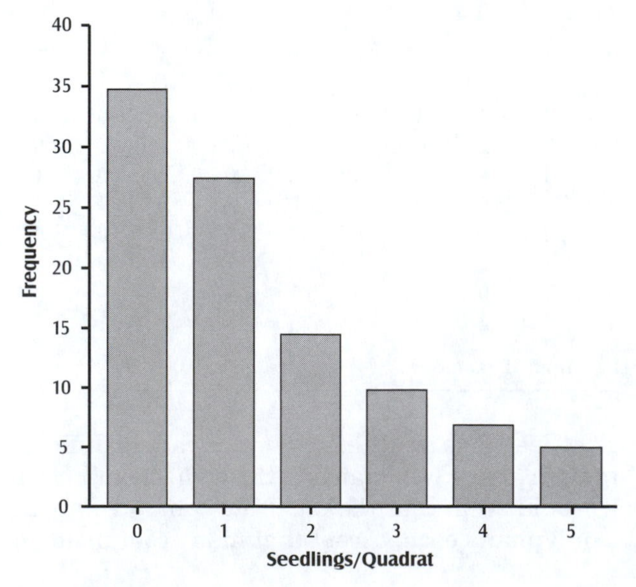


Data from M. Hamas

## 3.2 Frequency Distributions of Continuous Variables

You will recall that continuous variables may assume any imaginable value between certain limits. Accordingly, their graphic and tabular presentations are somewhat different from those of discrete variables. To create these, we need to pool together values that fall within specific class limits (sometimes called "bins").

### • Example 3.2
Frequency distribution of a continuous variable

A simple random sample of 172 adult male mosquito fish was collected from a large population, and the total length of each fish was measured to the nearest millimeter. These data are shown in table 3.3.

**Table 3.3** Total lengths (in mm) of 172 male mosquito fish

| | | | | | | | | | |
|---|---|---|---|---|---|---|---|---|---|
| 30 | 22 | 23 | 28 | 19 | 22 | 21 | 23 | 22 | 24 |
| 17 | 22 | 24 | 20 | 20 | 23 | 20 | 18 | 21 | 25 |
| 19 | 22 | 21 | 20 | 20 | 23 | 22 | 18 | 22 | 21 |
| 22 | 24 | 17 | 22 | 23 | 21 | 21 | 19 | 24 | 20 |
| 19 | 22 | 22 | 21 | 23 | 20 | 21 | 20 | 20 | 23 |
| 22 | 20 | 20 | 24 | 24 | 24 | 18 | 19 | 21 | 21 |
| 24 | 19 | 22 | 23 | 24 | 28 | 22 | 24 | 21 | 20 |
| 23 | 23 | 22 | 22 | 21 | 21 | 22 | 20 | 21 | 24 |
| 23 | 21 | 25 | 25 | 24 | 25 | 23 | 23 | 22 | 20 |
| 19 | 22 | 23 | 22 | 24 | 24 | 19 | 20 | 24 | 24 |
| 22 | 23 | 25 | 22 | 20 | 20 | 24 | 20 | 23 | 22 |
| 22 | 21 | 20 | 21 | 24 | 20 | 24 | 21 | 21 | 22 |
| 19 | 22 | 23 | 22 | 22 | 22 | 20 | 23 | 22 | 22 |
| 22 | 21 | 23 | 23 | 20 | 21 | 21 | 20 | 21 | 22 |
| 19 | 22 | 20 | 20 | 21 | 19 | 24 | 22 | 21 | 20 |
| 20 | 22 | 22 | 23 | 20 | 19 | 24 | 22 | 21 | 21 |
| 23 | 29 | 21 | 21 | 21 | 21 | 24 | 23 | 24 | 22 |
| 24 | 25 | | | | | | | | |

Once again, inspection of the data in this form makes interpretation difficult. Length is a continuous variable measured on a ratio scale. The information contained in a frequency distribution of such a variable is somewhat different from that contained in a frequency distribution of a discrete variable, as a continuous variable may assume any value within a certain range, while a discrete variable may not. Thus, when constructing a frequency distribution of a continuous variable, it is customary to group measurements into classes containing only individuals that fall within a certain range of the variable under consideration. For example, we might group all of the male mosquito-fish lengths in table 3.3 that measured 17 mm and 18 mm into one class, those that measured 19 mm and 20 mm into another, and so on. Keep in mind that measurement was rounded to the nearest millimeter, so that, in effect, 17 mm really includes individuals from 16.5 mm to 17.5 mm, and so on. Thus, our 17-mm-to-18-mm class actually includes individuals from 16.5 mm to 18.5 mm (a range of 2 mm) and our 19-mm-to-20-mm class includes individuals from 18.5 mm to 20.5 mm (also a range of 2 mm). In this case we are using a class interval of 2 mm. The choice of class interval is somewhat arbitrary and depends on the number of measurements and on the range of their values. Using too many classes tends to be confusing, while using too few conveys too little information. Trial and error is not a bad way of arriving at a satisfactory class interval! Once a class interval is chosen, a value halfway between the smaller and the larger limit of each class is designated as the class mark. For example, in the 16.5-mm-to-18.5-mm class above, 17.5 mm is the class mark. The class mark for the 18.5-mm-to-20.5-mm class interval is 19.5 mm, and so on.

Table 3.4 is a frequency distribution of the data in table 3.3, using a class interval of 2 mm. The smallest fish in the sample was 17 mm (actually 16.5 mm to 17.5 mm), which tells us that the first class includes all individuals that were between 16.5 mm and 18.5 mm. In other words, it includes all of those individuals that had a measured length of 17 mm or 18 mm. The second class includes individuals between 18.5 mm and 20.5 mm, and so on. Notice that the sum of the frequencies is 172, the number of fish that were measured (table 3.4).

We implied earlier that creating a frequency distribution by hand is a bit of an art. However, there are some guidelines we can follow. The class interval is simply the range of the data divided by the number of classes. The number of classes depends on the sample size. For example, with 100 measurements, we can usually manage about 10 classes. So, if our data ranged in value from 50 to 350 (a range of 300), the interval width would be 300 ÷ 10 = 30. We would then make the first bin 49.5–79.5, the second bin 79.5–109.5, and so on.

## 3.3 Histograms and Their Interpretation

Consider again the frequency distribution displayed in table 3.4. This table is useful for organizing data and doing calculations, but is not so easy to visualize. Here a graphic is helpful. Figure 3.2 shows a histogram of these data. Note that in the distribution in table 3.4 and in the histogram in figure 3.2 there is no space between adjacent classes. This reflects the basic nature of a continuous variable. In contrast, the bar graph shown earlier has spaces between the classes, consistent with the nature of a discrete variable.

Frequency distributions plotted as histograms are very useful for data inspection. Often, we would like to know if the distribution is symmetric or skewed (where the peak is shifted to the left or the right). For example, the mosquito-fish lengths (fig. 3.2) clearly have a generally symmetric distribution. Such data inspections become particularly useful when we check assumptions, such as normality (chapter 6). The histogram also allows us to visualize the value of

**Table 3.4** Total length of male mosquito fish (in mm)

| Measured Length | Implied Length | Class Mark ($x$) | Frequency ($f$) | ($fx$) | Cumulative Frequency |
|---|---|---|---|---|---|
| 17–18 | 16.5–18.5 | 17.5 | 5 | 87.5 | 5 |
| 19–20 | 18.5–20.5 | 19.5 | 40 | 780.0 | 45 |
| 21–22 | 20.5–22.5 | 21.5 | 70 | 1505.0 | 115 |
| 23–24 | 22.5–24.5 | 23.5 | 47 | 1104.5 | 162 |
| 25–26 | 24.5–26.5 | 25.5 | 6 | 153.0 | 168 |
| 27–28 | 26.5–28.5 | 27.5 | 2 | 55.0 | 170 |
| 29–30 | 28.5–30.5 | 29.5 | 2 | 59.0 | 172 |
| | | SUM | 172 | 3744.0 | |

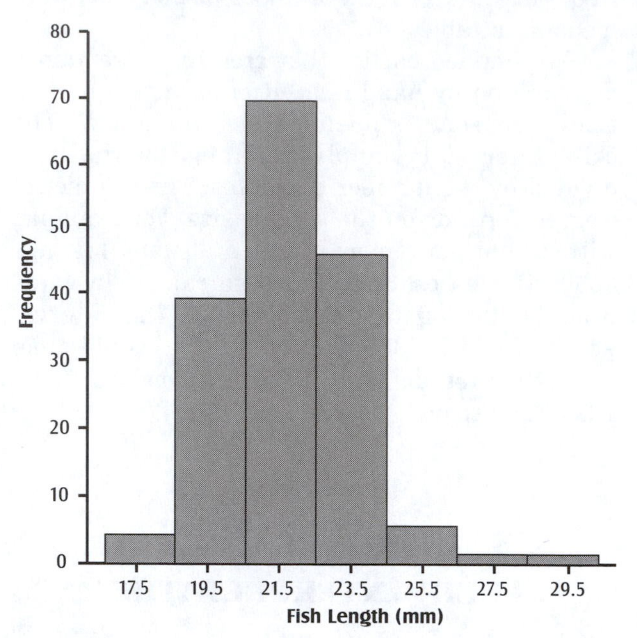


**Figure 3.2** Histogram of male mosquito-fish lengths, in mm

the mean (arithmetic average), as well as the amount of variation around the mean. For example, the mean length of mosquito fish in our example is around 21 or 22 mm. We will later calculate various descriptive statistics (chapter 4); using graphics helps us to quickly spot if our answers make sense.

Most statistical programs for computers have some graphing capabilities, including frequency distributions. (After you have organized and plotted a couple of histograms by hand, you will appreciate the labor saving!) Figure 3.3 is a histogram generated by MINITAB using a large data set.

Occasionally, frequency distributions of a continuous variable are displayed with a line graph. Rather than using bars, points are plotted on a graph with class marks on the $x$-axis and frequency on the $y$-axis. Such a graph is called a **frequency polygon**, and an example is shown in figure 3.4 using the data from table 3.4. Frequency polygons should not be used to show the frequency distribution of a discrete variable since the line "connecting the dots" indicates that the variable is continuous.

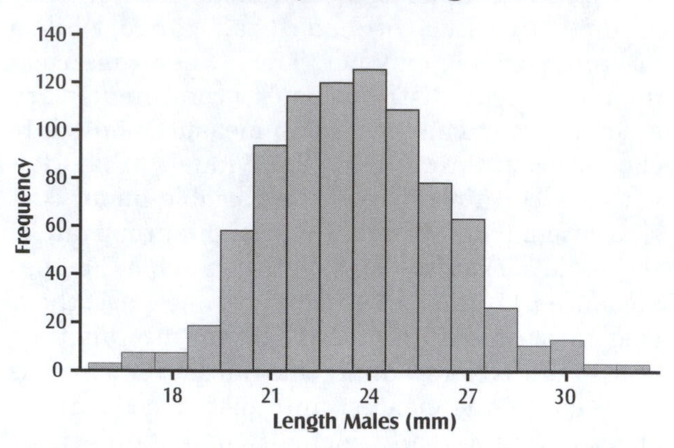


**Figure 3.3** MINITAB-generated histogram of 854 male mosquito-fish lengths, in mm

Data from digital appendix 2

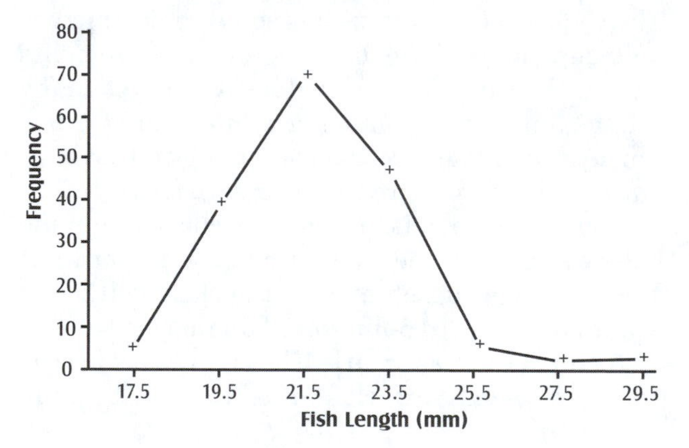


**Figure 3.4** Frequency distribution of male mosquito-fish length, in mm (an example of a frequency polygon)

## 3.4 Cumulative Frequency Distributions

Occasionally, we prefer to display data as cumulative frequencies, the number of observations in a particular class plus all lower classes. An example is shown for the mosquito-fish data in table 3.4 and displayed in figure 3.5. Notice that the height of the bars do not increase very much beyond the 23.5-mm size class, revealed further by the flattening of the frequency polygon curve. Cumulative frequencies can also be expressed as percents and used to answer such questions as, "What % of the population is smaller than or equal to size $x$?" This type of question is frequently raised with student scores from standardized exams (e.g., "What percentile was your score on the SAT [or ACT] exam?"). Cumulative frequency polygons also are useful for revealing diminishing returns from extensive sampling; e.g., "How many samples are enough to detect most of the species?".

**Figure 3.5** Cumulative frequency distribution of mosquito-fish lengths as histogram and frequency polygon

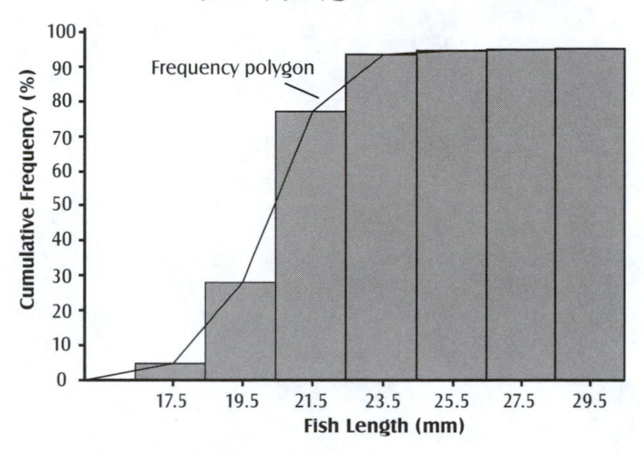


Adapted from J. Sumich, unpublished.

## 3.5 Other Handy Graph Types

A wide variety of graphics can be used to display different types of numerical data. We can compare the differences between groups by showing their averages with bars (or dots) and lines illustrating amounts of variation (e.g., the standard error of the mean, chapter 7). These illustrations are handy displays to support results from formal hypothesis tests, such as t-tests (chapter 9) or analysis of variance (chapter 10).

The relationship between two continuous variables is typically displayed with an *x-y* graph, also called a **scatterplot**. Examples of such graphs are displayed in chapters 12 and 13. They provide supporting evidence for formal statistical tests, such as regression analysis and correlation analysis. Students who have created a standard curve in chemistry have done such plots.

The proportion of individuals belonging to different categories (attributes) is commonly analyzed with chi-square tests (chapter 14). Such data are usually organized in tables, but can also be nicely displayed with a bar graph. A pie chart may also be used. Both of these graph types are commonly seen in the media and in published scientific papers.

## Key Terms

- bar graph
- bins (classes)
- cumulative frequency distribution
- frequency
- frequency polygon
- histogram
- frequency distribution
- scatterplot

## Exercises

Answer the first three questions by hand and the remaining using a computer statistical package.

3.1 Below are the heights (in cm) of a sample of 148 male humans. Construct a frequency distribution table and a histogram of these data. A class interval of 2 cm is suggested.

187 171 181 180 178 171 174 177 172 178 182
187 176 179 190 185 192 184 182 178 187 173
185 184 184 183 185 197 202 181 181 191 178
187 185 186 174 174 182 195 182 180 182 182
179 183 178 185 178 190 180 175 169 176 182
185 179 180 187 178 170 181 200 161 181 173
178 182 181 181 181 172 185 188 188 177 176
173 174 176 189 180 182 188 184 179 177 177
183 196 184 173 180 180 180 184 175 176 186
187 182 187 174 178 191 182 174 178 191 178
173 183 191 191 180 187 184 177 186 194 185
189 193 189 192 189 181 177 176 190 173 179
180 184 176 180 178 171 182 173 184 193 182
185 178 190 190 183

3.2 Given below are the lengths (in millimeters) of a sample of 100 largemouth bass. Construct a frequency distribution table and histogram of these data.

210 325 285 402 350 240 409 330 295 325
241 383 361 355 200 432 130 114 170 135
371 307 207 175 177 261 166 376 216 152
347 322 387 233 284 394 297 321 281 66
90 115 250 201 175 320 370 312 370 320
175 201 250 115 95 70 289 312 322 258
188 192 350 200 199 180 190 180 200 200
349 192 189 260 320 432 456 331 418 357
304 316 336 368 415 370 336 315 305 420
310 397 193 394 199 338 296 312 269 203

3.3 Some species of chironomid larvae inhabit the leaves of pitcher plants. The number of larvae per leaf in a random sample of 197 leaves were as follows. Construct a bar graph of these data. What is the total number of chironomid larvae found in these 197 leaves?

| Larvae/ Leaf ($x$) | Number of Leaves Containing $x$ Number of Larvae |
|---|---|
| 0 | 10 |
| 1 | 15 |
| 2 | 27 |
| 3 | 18 |
| 4 | 38 |
| 5 | 57 |
| 6 | 22 |
| 7 | 5 |
| 8 | 2 |
| 9 | 3 |
| 10 | 0 |

3.4 Using the data on the lengths of bluegill sunfish in digital appendix 1 ("bluegill"), construct a histogram of this variable. A 10-mm class interval is recommended.

3.5 Using the data on the length of female mosquito fish in digital appendix 2 ("mosquitofish"), construct a histogram of this variable. A class interval of 4 mm is suggested.

3.6 Using the data on pulse rate in digital appendix 3 ("StudentData"), construct a histogram of this variable (for males and females).

3.7 Using the data on pulse rate in digital appendix 3, construct a histogram of the pulse rate of females.

3.8 Using the data on pulse rate in digital appendix 3, construct a histogram of the pulse rate of males.

3.9 Locate an article in a field of your interest that contains numerical data. How are the data summarized for display in tables and/or figures? Can you find an example of a frequency distribution?

# 4

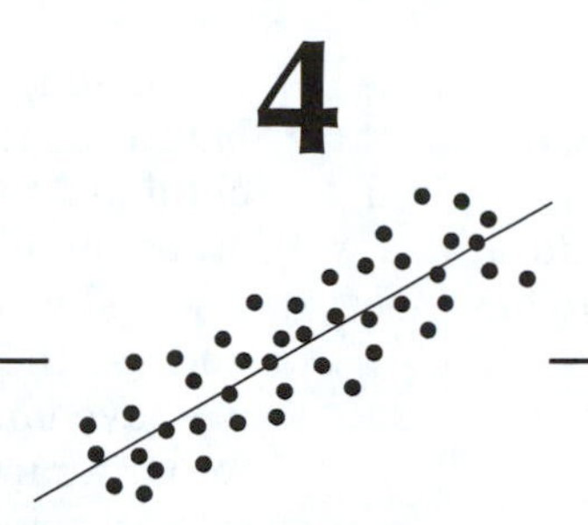

# Descriptive Statistics

## Measures of Central Tendency and Dispersion

If we measure some characteristic on a number of individuals in a population, we often find that the measurement differs among individuals. We may then discover that there is a certain value around which our measurements tend to cluster. Statistics gives us ways to describe this central tendency and to describe the variation of individuals from this point. In this chapter we consider several ways to describe the central tendency and the variation around it.

## 4.1 Sample Statistics and Population Parameters

It would be quite unusual indeed to have all of a population available for measurement. Even if we did, we might not want to take all the trouble to measure so many things. Accordingly, we calculate such things as central tendencies and variation using a random sample from the population of interest and use them to estimate the population's values. These values in the population as a whole are called **parameters**, and they are usually unknown. In a sample they are called **statistics** (plural; the singular is statistic). Later in this chapter, we calculate several statistics. Three of these are shown below with their accompanying population parameters. The true mean of a population is a parameter, and the mean of a sample taken from that population is a statistic. Parameters are conventionally symbolized by Greek letters and statistics by Roman letters, in this manner:

| | parameter | statistic |
|---|---|---|
| mean | $\mu$ | $\bar{x}$ |
| median | $\theta$ | $M$ |
| variance | $\sigma^2$ | $s^2$ |
| standard deviation | $\sigma$ | $s$ |

The Greek letters are: mu ($\mu$), theta ($\theta$), and sigma ($\sigma$).

## 4.2 Measures of Central Tendency

Several values may be used to describe the **central tendency** of a sample or a population. Which is most appropriate depends on the scale of measurement used and on the information we wish to convey.

### 4.2.1 The Mode

With data measured on a nominal scale, the only measure of central tendency is the **mode**, the most frequent item in the data set. For example, in a group of 15 red marbles, 10 white marbles, and 5 blue marbles, the mode is red marbles. The mode can be applied to data measured on other scales and is also used to describe the peak in a frequency distribution.

### 4.2.2 The Median ($\theta$, $M$)

The median is a useful measure of central tendency when data are measured on at least an ordinal scale. The items are arranged in order from smallest to largest. The **median** is the value that has an equal number of items above and below it. For example, suppose we have the following ordinal data on the conditions of 15 patients at Hopeless Hospital:

| condition | near death | deteriorating | stable | improving |
|---|---|---|---|---|
| frequency | 2 | 8 | 4 | 1 |

The median condition is "deteriorating". This result is made more clear if we collapse the frequency distribution into an ordered list of codes for the patient conditions, with the middle condition (median) underlined:

N N D D D D D <u>D</u> D D S S S S I

On an interval or ratio scale, the median is the center of an ordered array of numbers. For example, in the measurements

2, 2, 2, 3, 3, 4, 4

3 is the median, since this value has an equal number of measurements above and below it. When there are an even number of observations, none can have an equal number above and below. In this case the median is the value halfway between the two central values. In the series

2, 2, 2, 3, 4, 4, 4, 4

the median is 3.5 (the average of 3 and 4). Note that in the series

1, 2, 2, 3, 4, 10, 100, 1000

the median is still 3.5, even though the latter three numbers have a great deal more spread and cover a much larger range than the first three numbers. The median is a very useful measure of central tendency for both rank and measurement data.

### 4.2.3 The Mean ($\mu$, $\bar{x}$)

The **mean** is an arithmetic average and is the most common measurement of central tendency when data are measured on an interval or ratio scale. If we measure all of the individuals in a population of interest and compute a mean based on these measurements, we have in fact obtained the true population mean ($\mu$). On the other hand, if we measure only a randomly selected sample of the members of the population and use these measurements to compute a mean, we have obtained a sample mean, symbolized by $\bar{x}$. Ordinarily, we deal with samples, so we calculate sample means.

If $x_1, x_2, x_3, \ldots, x_n$ are individual measurements from a sample of size $n$, their mean ($\bar{x}$) is:

$$\bar{x} = \frac{\sum x}{n} \tag{4.1}$$

or, in words, the sum of all of the $x$ values divided by the total number of $x$ values. (By the way, the symbol $\Sigma$ is the capitalized version of the Greek letter "sigma," and signifies summation.) The sample mean ($\bar{x}$) is an estimate of the true population mean ($\mu$), which we ordinarily do not know. If we did measure all items in the population, $\mu$ would be calculated by the same formula, replacing $\bar{x}$ with $\mu$.

### 4.2.4 Positions of Mean, Median, and Mode in Symmetric and Skewed Distributions

Suppose we collect a large number of observations and construct a frequency distribution. Some distributions are symmetric, with an equal spread of data on either side of the peak in frequency (mode). When this occurs, the mean, median, and mode all occur at the same value of $x$. Other distributions are skewed, with a peak on one side or the other of the distribution and a long "tail" to the other side. In a skewed distribution, the mode is again (by definition) the value of $x$ where the peak in frequency occurs. The mean is located farther out toward the long tail and the median occurs between the mean and the mode. Examples of symmetric and skewed distributions are shown in figure 4.1.

Rare observations in the tail are sometimes called **outliers**. They heavily influence the mean. Let's imagine a common situation with a skewed distribution: incomes of workers. To simplify our problem, we will use a small sample.

**Figure 4.1** Symmetric and right-skewed frequency distributions

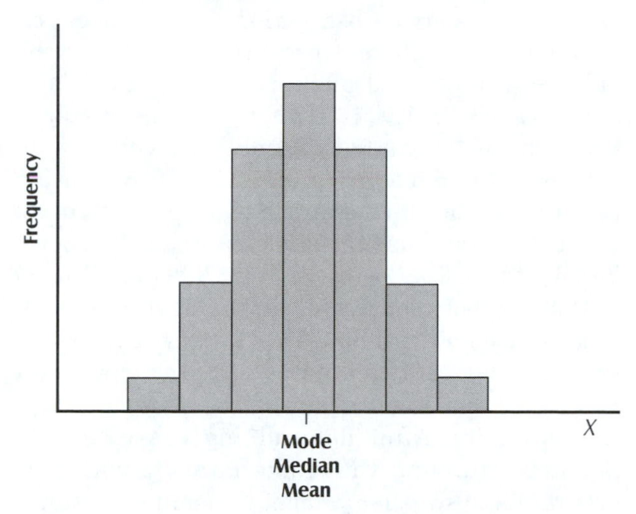


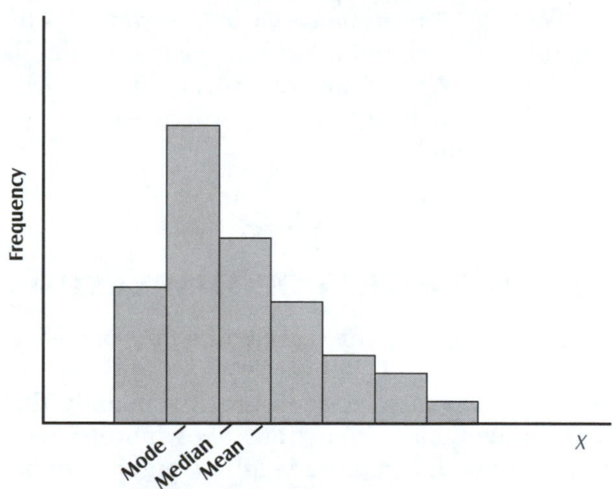


Adapted from T. H. Wonnacott and R. J. Wonnacott, *Introductory Statistics for Business and Economics.* © 1977. Reprinted with permission of John Wiley & Sons, Inc.

## • Example 4.1

Annual incomes of the 9 employees of Redundancy Manufacturing (in $1000s): 15 17 21 21 24 27 27 31 225.

With so few measurements, the mode is not very meaningful. However, we can determine the mean and median. Because of the high salary of one person (presumably the boss), the mean is inflated to $45,333 (408/9), a value considerably larger than any employee other than the boss. Indeed, if we omitted this outlier, the mean salary of all the others is $22,875. The median salary for this company is $24,000. Notice that this value remains the same even if the boss took a large pay cut!

### 4.2.5 Weighted Mean

The mean is typically calculated as the arithmetic average, as shown above. However, there are occasions when other types of means make more sense. Most common is the **weighted mean**, calculated from data arranged in a frequency distribution as follows:

$$\bar{x} = \frac{\Sigma fx}{\Sigma f} \tag{4.2}$$

where $\Sigma fx$ is the sum of products between values of $x$ and their frequencies (weights), and $\Sigma f$ is the sum of the frequencies, also equal to $n$, the total sample size.

## • Example 4.2

Most undergraduates are concerned with their grade point averages, which are simply weighted means! For example, let's consider Sue Sigma, who has taken 50 credits so far and passed 46 of these. When all her grades are compiled into a frequency distribution (look familiar?), we get:

| grade | grade value $x$ | number of credits attempted $f$ | grade points $fx$ |
|---|---|---|---|
| A | 4 | 12 | 48 |
| B | 3 | 18 | 54 |
| C | 2 | 12 | 24 |
| D | 1 | 4 | 4 |
| F | 0 | 4 | 0 |
| | | $\Sigma f = 50$ | $\Sigma fx = 130$ |

$$\text{GPA} = \bar{x} = \frac{\Sigma fx}{\Sigma f} = \frac{130}{50} = 2.60$$

Weighted means are calculated when different measures have different levels of importance (weights). In the GPA example, the number of credits represents the weight. As another example, consider a problem from environmental toxicology. Suppose we had PCB measurements ($x$) of tissues taken from different individual lake trout, which we had previously been weighed. In order to estimate the overall PCB levels in lake trout, we would determine a weighted mean, using the fish weight as the "weight" ($f$ above).

Weighted means are also useful to determine an average from data that already are arranged in a frequency distribution. For instance, consider the data on mosquito-fish lengths in the last chapter (table 3.4). The data are tabulated for us. The weighted mean would then be $\Sigma fx \div \Sigma f$

= 3744 ÷ 172 = 21.8 mm. Notice that this value is near the center of the frequency distribution (fig. 3.2). Calculating a mean from such a frequency distribution is often useful to get additional information from summary data in a publication. Of course, if we had the original data at our disposal, we could simply calculate the sample mean and have no need to calculate the mean from the frequency distribution.

### 4.2.6 Other Measures of Location

Recall that the median is a measure of central location. In an ordered array, the median divides the top half of observations from the bottom half. There are other locations we could mark. Having taken standardized tests, most of us are familiar with **percentiles**, which divide up an array into 100 slices. If you scored in the 90th percentile in the verbal test, this means that 90% of the people taking the test scored below you. The median is identical to the 50th percentile. There are occasions when we want to divide the distribution into quarters (e.g., see interquartile range below). The **first quartile** (Q1) divides the bottom quarter from the top three quarters and is the same as the 25th percentile. The median is also called the **second quartile** (and 50th percentile), as it divides the bottom two quarters from the top two quarters. The **third quartile** (Q3) divides the top quarter from the bottom three quarters.

---

#### • Example 4.3

Consider an imaginary example from Wisconsin. A farmer has 14 cows, from which he determined milk yields. Milk yield is measured in pounds per day and the cows' measurements are sorted in ascending order:

19 23 26 30 32 34 37 37 39 41 44 44 46 55

---

Using equation 4.1, we calculate the mean milk yield to be 36.2 pounds per day. Since the number of measurements is even, the median is the average between the 7th and 8th positions (both 37). A "quick and dirty" way to find the first quartile (Q1) is to find the number that divides the lower half of observations in half. In this example, that number occurs at about the 4th position, where the value is 30. This number is actually slightly off, as the lower 25% of observations is cut off at a number slightly less than this number. The exact approach requires some interpolation:

1) Calculate (n + 1)/4.

2) If (n + 1)/4 is an integer (whole number), then Q1 is the value of this integer.

3) If (n + 1)/4 is not an integer, then Q1 is a value between two observations and its location between the numbers is proportional to the to the value of the decimal answer. For the cows data, (14 + 1)/4 = 3.75. Q1 is thus ¾ of the distance between the 3rd and 4th numbers, a value of 29. (This was figured as: 26 + 0.75(30 – 26) = 26 + 3 = 29.) To find the third quartile (Q3), first calculate 3 × (n + 1)/4 and use the same approach as above. For the cows, 3 × (14 + 1)/4 = 11.25. Q3 is thus ¼ of the way between the 11th and 12th numbers, in this case both 44. So Q3 = 44. Save these numbers for later. Most computer statistical packages determine quartiles as part of displaying descriptive statistics. After determining the values by hand, try running the same data set with this program and see that you get the same answer.

We have determined a variety of statistics for central tendency and other locations in a sample distribution. We are now ready to consider variation, the stuff that makes statistics necessary and interesting!

## 4.3 Measures of Dispersion

It is unlikely that individuals of a biological population will all be exactly alike in any variable that one cares to measure. We usually find that individuals differ from one another with respect to such things as length, weight, number of warts, and so on. As variation is the usual state of the natural world, it is quite important to measure it. Controlling blood pressure with medication allows not only a reduction in average diastolic blood pressure, but also its swings from high to low. A drunken driver on the highway is dangerous to others not so much because of their mean position, but because of their variability in position—weaving down the highway! For all these things, there is some value about which individual measurements tend to cluster. When measuring with an interval or ratio scale, this value is usually the arithmetic mean, and when measuring with an ordinal scale this value is the median.

Consider the data on length of male mosquito fish. Refer back to chapter 3 to see the raw data (table 3.3) and histogram from these data (figure 3.2). The mean for this sample is 21.83 mm. Note that the most frequently occurring size class is made up of individuals whose size clusters about the mean value. Note also that there are progressively fewer individuals in each size class as we move from the mean toward either smaller or larger individuals. Frequently, we need to measure this spread of the individual observations around the mean. The terms spread and **dispersion** both refer to the same thing, the variation in observations around the mean (or median).

## 4.3.1 The Range

The **range** is the difference between the largest and smallest items in the sample. The range is expressed in the same units as the original measurement. For example, the smallest measurement in the data on male mosquito-fish length (table 3.3) is 17 mm, and the largest is 30 mm. The range is therefore 30 mm – 17 mm, or 13 mm. Similarly, for the Wisconsin cow data above, the range is 55 – 19, or 36 pounds per day.

The range is strongly influenced by even a single extreme value. Suppose, for instance, that a single cow produced 162 pounds per day. The range would now be 143 pounds per day, even though the spread among the other cows was the same. The range is thus limited as a description of the amount by which any individual measurement is likely to vary from the mean. Nevertheless, reporting the maximum and minimum values is a common description of environmental and physiological data, because of the tolerance limits of organisms. For instance, your body temperature may usually vary little from 98.6° F, but one instance of extremely high temperature, say over 107°F, can be fatal. When reporting the range, the best approach is simply to report both the minimum and maximum and the reader can do what they want with these numbers.

## 4.3.2 Interquartile Range

Because of the extreme effect of outliers on the range, another measure of dispersion is often presented. The **interquartile range** is the difference between the third and first quartiles and represents the spread of the middle 50% of the (ordered) values. For the cow data, the interquartile range is about 44 – 29, or 15 pounds per day. In other words, the middle half of cows showed a range of 15 pounds per day. The interquartile range often appears on a graph. The **Box-Whisker Plot** (usually just called the Box plot) illustrates the locations of the median, interquartile range, and range, and is often used to compare these characteristics in different groups (figure 4.2). For this example, we have added another hypothetical sample, one from Missouri, which has some very productive cows!

**Figure 4.2** Box plots of milk yields data from two locations, as generated by MINITAB

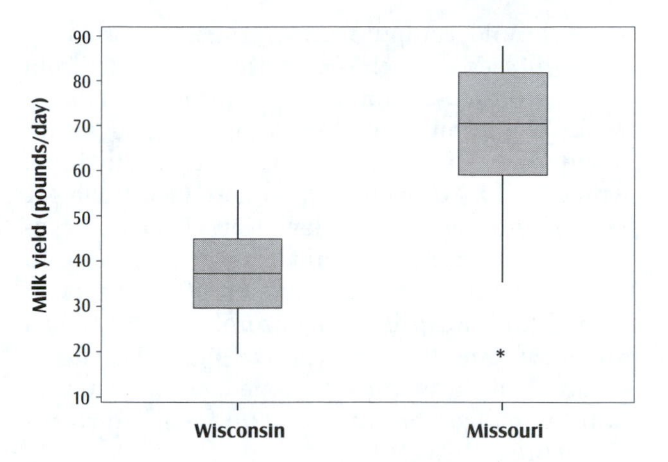


The median is shown as the horizontal line inside the box and the first and third quartiles are at the boundaries of the box. The vertical lines (called "whiskers") represent the range of most values. Specifically, the top whisker extends up from Q3 to the maximum value that is still within the limit of Q3 + 1.5 × (Q3 – Q1) and the bottom whisker extends down to the minimum value which is still within the limit Q1 – 1.5 × (Q3 – Q1). Outliers show up as an *. (One outlier is shown for the Missouri data.) Since computer software can readily generate such plots, most users don't need to do these calculations.

## 4.3.3 Standard Deviation ($\sigma$, $s$)

Next, let's consider a widely used measure of variation that takes into account all of the observations in a distribution. The **standard deviation** is, in effect, a measure of the average amount by which each observation in a series of observations differs from the mean. We will use an imaginary set of data to illustrate the calculation and meaning of the standard deviation. We assume that the measurements were made on an interval or ratio scale, since the following operations cannot be performed on ordinal or nominal measurements. We also assume, as we often do, that we have a random sample from a population. The true population standard deviation, denoted by the Greek symbol sigma ($\sigma$), is usually unknown and must be estimated from our sample, calculating the sample standard deviation ($s$).

Suppose we collect a sample of 8 measurements: 1, 2, 2, 3, 3, 4, 4, 5. The sample mean for these 8 measurements is:

$$\bar{x} = \frac{1+2+2+3+3+4+4+5}{8} = \frac{24}{8} = 3.0$$

We could compute an average deviation by first subtracting each observation (variant) from the mean, then summing these values and dividing by the number of observations. However, we would find that the sum of these **deviations**, symbolized by $d$, would equal zero (see table 4.1 below), and the average deviation of the observations from the mean would therefore be zero. In fact, we would find that the sum of the deviates around any mean is always zero! Squaring each deviate eliminates the negative signs. When the squared deviates are summed, the resulting value is called the **sum of squares**, which we will abbreviate as SS.

**Caution**

The term "sum of squares" (abbreviated SS) is commonly used in this book and refers to the sum of squared deviations from a mean. The sum of squares does not mean simply the sum of the squared observations ($\Sigma x^2$ below). This is a very common mistake by students.

The sum of squares (SS) is obtained by squaring the difference (deviation) between each observation and the mean of all of the observations, and then summing the squared differences, or

$$SS = \Sigma(x - \bar{x})^2 \qquad (4.3)$$

As we shall see later, the SS is a very important quantity in many statistical procedures. In our example, the sum of squares is 12 (table 4.1).

The procedure of subtracting each observation from the mean, squaring the difference, and summing these squared deviates is computationally tedious if more than a few observations are involved. A much more efficient method of computing the sum of squares is by formula 4.4:

$$SS = \Sigma x^2 - \frac{(\Sigma x)^2}{n} \qquad (4.4)$$

The first term in the equation, $x^2$, instructs one to square each observation and then sum the results. The second term, $(\Sigma x)^2/n$, instructs one to sum the observations, square the sum, and then divide that result by the number of observations.

If the sum of squares is divided by the number of observations minus one (also called the **degrees of freedom**), another important value, called the **variance**, is obtained:

$$s^2 = \text{variance} = \frac{SS}{n-1} \qquad (4.5)$$

The variance represents the average of the squared deviations in the sample. If the original measurements were in mm, the units of the variance would be mm$^2$. Ordinarily, we desire a statistic with the same scale and units as our original measurements. To rectify this condition, we need only take the square root of the variance. This value is called the **standard deviation**. The standard deviation has the same units as the original measurements.

$$s = \text{standard deviation} = \sqrt{\frac{SS}{n-1}} = \sqrt{\text{variance}} \qquad (4.6)$$

In our example above, the standard deviation is 1.309. Work through this example, summarized in table 4.1, to make certain that you thoroughly comprehend these concepts.

**Table 4.1** Calculation of mean, variance, and standard deviation from a sample

The sample mean equals 3.0 and sample size is 8.

| $x$ | $d = x - \bar{x}$ | $d^2$ |
|---|---|---|
| 1 | −2 | 4 |
| 2 | −1 | 1 |
| 2 | −1 | 1 |
| 3 | 0 | 0 |
| 3 | 0 | 0 |
| 4 | 1 | 1 |
| 4 | 1 | 1 |
| 5 | 2 | 4 |
| | $\Sigma = 0$ | SS = 12 |

$s^2$ = variance = 12/7 = 1.7143

$s$ = standard deviation = $\sqrt{1.7143} = 1.3093 \approx 1.3$

Notice that we have carried all figures in this example and rounded only at the end. We round these sample statistics to one more significant figure than the original measurements. So, after rounding, the standard deviation should be reported as 1.3 and the mean as 3.0.

**Table 4.2** Descriptive statistics from MINITAB

A) Milk yields from Wisconsin cows (example 4.3)

**Descriptive Statistics: MilkYield**

| Variable | N | N* | Mean | SE Mean | StDev | Minimum | Q1 | Median | Q3 |
|---|---|---|---|---|---|---|---|---|---|
| MilkYield | 14 | 0 | 36.21 | 2.61 | 9.76 | 19.00 | 29.00 | 37.00 | 44.00 |

| Variable | Maximum |
|---|---|
| MilkYield | 55.00 |

B) Lengths of male mosquito fish (example 3.2)

**Descriptive Statistics: MosquitoFishL**

| Variable | N | N* | Mean | SE Mean | StDev | Minimum | Q1 | Median | Q3 |
|---|---|---|---|---|---|---|---|---|---|
| MosquitoFishL | 172 | 0 | 21.831 | 0.155 | 2.026 | 17.000 | 20.000 | 22.000 | 23.000 |

| Variable | Maximum |
|---|---|
| MosquitoFishL | 30.000 |

**Caution**

Beware of rounding errors. When doing calculations, always carry all the figures in your calculator memories. Round only at the final step. Rounding errors can sometimes be quite significant! Clever embezzlers have occasionally taken advantage of rounding errors to steal money from banks.

### 4.3.4 Coefficient of Variation

There are times when we wish to re-express the standard deviation in such a way as to allow a comparison between groups that are very different in some way. Suppose, for instance, that we wanted to compare the variability in body weights between mice and elephants. Simply comparing their standard deviations just doesn't make sense; after all, these animals would have been measured on different scales. Instead, we'll standardize their standard deviation relative to their mean body weight. To do so we compute the **coefficient of variation** as:

$$CV = \frac{s}{\bar{x}} \times 100\% \qquad (4.7)$$

The following hypothetical example will illustrate:

**• Example 4.4**

| | Body weight (kg) | |
|---|---|---|
| | mice | elephants |
| $\bar{x}$ | 0.0125 | 1,240 |
| s | 0.0072 | 625 |
| CV (%) | 57.6 | 50.4 |

The example shows that, after standardizing to their mean body weights, variability in body weights is quite similar for mice and elephants.

## 4.4 Descriptive Statistics from a Computer

Virtually all statistical programs for computers give summary statistics for a sample variable. The summary statistics from MINITAB are shown in table 4.2 for two data sets. In these printouts, most of the statistics should look familiar, with a few exceptions. "N*" is the number of missing observations. "SE Mean" is the "standard error of the mean," a measure of variation which we consider in a later chapter.

## 4.5 Visualizing the Location of the Mean and Standard Deviation

In chapter 3, we saw the importance of organizing data into frequency distributions and in this chapter learned methods for computing various descriptive statistics. Before moving on, let's briefly think about these statistics by superimposing them on a frequency distribution. Refer to figure 4.3, which shows the histogram for male mosquito fish seen earlier (figure 3.2), but now with some statistics calculated from the raw data. Notice that the interquartile range encompasses the middle half of the observations (think area of the bars). Also notice how far out on the distribution the standard deviations extend. In chapter 6, we see that in a normal distribution about 95% of the observations should fall within ± 2 SD units from the mean. More on that later!

**Figure 4.3** Histogram of male mosquito-fish lengths, using class intervals of 1mm. Descriptive statistics for variation are shown.

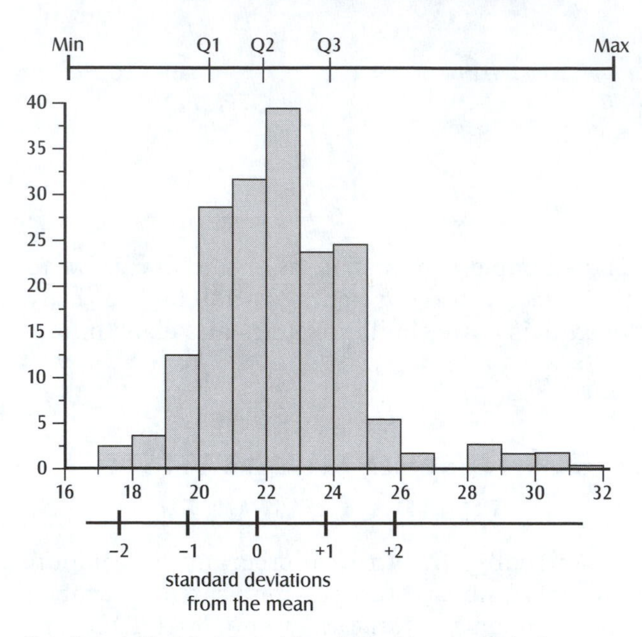


Data from table 3.3. Adapted from J. Sumich, unpublished.

## Key Terms

| | |
|---|---|
| Box-Whisker Plot | outlier |
| central tendency | parameter |
| coefficient of variation | range |
| dispersion | standard deviation |
| interquartile range | sum of squares (SS) |
| mean | variance |
| median | weighted mean |
| mode | |

## Exercises

4.1 For the following sample, compute the following statistics using a calculator: mean, median, range, interquartile range, variance, standard deviation, coefficient of variation. What is the sample size? Show all steps in your solutions. Note that you will need to rewrite the numbers in order from smallest to largest.

2 5 3 7 8 3 9 3 10 4 7 4 6 11 9
9 11 5 7 3 8 9 2 1 3 8 3 8 9 3

4.2 For the following sample, compute the same statistics as in exercise 4.1.

23 43 12 56 43 23 56 43 23 32 12 14 15

4.3 A random sample of 42 belted kingfishers was collected from various locations in North America and their culmen (bill) lengths were measured in mm:

| | | | | | |
|---|---|---|---|---|---|
| 48.1 | 50.8 | 48.8 | 56.8 | 57.7 | 47.0 |
| 56.8 | 60.2 | 55.8 | 59.2 | 52.5 | 50.4 |
| 48.0 | 57.1 | 51.8 | 52.3 | 47.8 | 58.0 |
| 53.4 | 55.2 | 51.0 | 59.3 | 61.5 | 61.2 |
| 57.8 | 50.1 | 56.0 | 56.5 | 55.8 | 56.5 |
| 56.3 | 59.8 | 61.8 | 56.2 | 57.5 | 59.3 |
| 62.4 | 61.1 | 59.9 | 55.6 | 56.8 | 59.2 |

Compute the same statistics as in exercise 4.1. Next, use a statistical package to calculate these same descriptive statistics. Compare your answers; if you detect a discrepancy, be sure to find out why.

4.4 Using the frequency distribution table you generated for the sample of 148 male heights (exercise 3.1), compute a weighted mean (eq. 4.2). Compare to the ordinary sample mean (eq. 4.1).

4.5 Using the data from exercise 3.2, compute the weighted mean for the largemouth bass lengths. Compare to the ordinary sample mean.

4.6 Using samples of 10, 20, and 30 bluegill-sunfish standard lengths from exercise 1.1, compute the mean and standard deviation of each sample. Do both by hand and with a computer. Consider the measurements in digital appendix 1 to be the entire population of interest, with a population mean of 120.03 mm and a population standard deviation of 41.95 mm. Are the sample means and standard deviations of your samples the same as the population mean and standard deviation? Why do you suppose that there might be a difference in the sample values and the population values? Save your calculated values for later use.

4.7 Using the samples of 10, 20, and 30 female mosquito-fish lengths from exercise 3.5, compute the mean and standard deviation for these samples. Do both by hand and with a computer. Consider the measurements in digital appendix 1 to be the entire population of interest, with a population mean of 34.29 mm and a population standard deviation of 5.49 mm. Are the means and standard deviations of your samples the same as the population mean and standard deviation? How do you account for the difference? Save your calculated values for later use.

4.8 Using the data from exercise 2.3, compute the mean number of ant lion pits/quadrat for this sample. Save your calculated value for later use.

4.9 Using the data from exercise 2.4, select and compute an appropriate measure of central tendency.

# 5

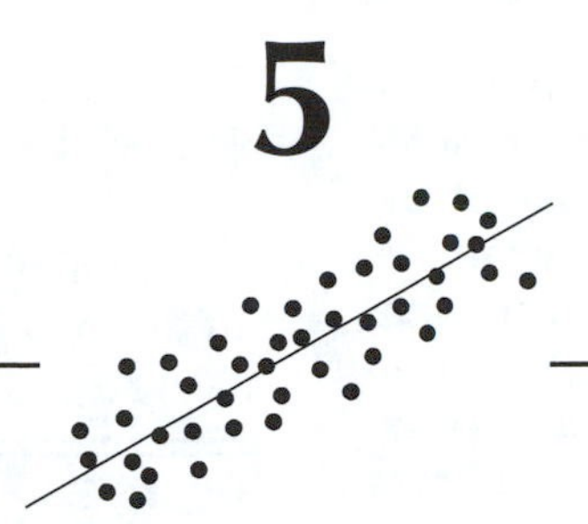

# Probability and Discrete Probability Distributions

Most events in life are uncertain. Will it rain today? Will the batter strike out? Will the phone ring tonight? **Probability** is the risk or chance that a particular event will occur in an uncertain world. Probability generates expectations that support informal decisions in such disparate fields as business, sports, and gambling. (Indeed, the formal rules of probability were developed in the 1600s by mathematicians employed by gamblers; now they are employed by insurance companies!). Probability rules are used in biology to predict the risk of genetic disease if we know a pattern of inheritance, and the risk of other diseases if we know some test outcome (using Bayesian statistics). Probability is a foundation of statistical inference (hypothesis testing), which we explore extensively later in this book.

In the current chapter we consider some of the basic rules of probability and some important discrete probability distributions. Rules of probability are introduced in the chapter as they are needed. In chapter 6, we examine another important probability distribution called the normal distribution.

## 5.1 Probability and Probability Distributions

The **probability** of an event is expressed numerically as a number between zero and 1 (or, in common usage, as a percent between zero and 100). A probability near 1 suggests that the event is very likely, while a probability near zero suggests that the event is very unlikely.

There are different ways of arriving at the probability of an event. One is based on a theoretical consideration (also called "classical probability"). Dice have 6 faces, each bearing a number of 1 through 6. Given that a die is "fair" (that is, equally as likely to land with one face up as any other), then the probability of obtaining any particular number on the toss of a single die is 1/6 or 0.1667. We don't need to toss any dice—the rules of the game determine the probability. Similarly, for genes following simple rules of Mendelian inheritance, we can predict the proportion of recessive phenotype following a monohybrid cross. In classical probability we know the "rules of the game" and thus can make predictions using some basic tricks, which we'll discover in this chapter.

A second important way of determining probability (empirical probability) is based on some prior knowledge of the relative frequency of the event in the population of interest. For example, the occurrence of children born with Down's syndrome (a form of mental retardation) is well known from birth records, so its incidence (relative frequency) can be calculated for different groups. From this information, we know that the incidence rises sharply in mothers after age 35. Doctors can thus give their patients an assessment of risk. Note that, in this case, we can arrive at this probability even though we don't know the precise mechanism. For both classical and empirical probability, probability was calculated from a ratio. This is the first of our probability rules, called the division rule.

**The Division Rule**

The probability of an event is the number of ways that an event can occur divided by the total number of events that may occur.

In the case of the dice, there was one way that a particular number could occur out of 6 possible occurrences.

If we toss a coin, there are 2 possible outcomes—a head or a tail. On any given toss, there is an equal chance of obtaining a head or a tail because the coin has an equal number of heads and tails—one of each—and because the 2 possible outcomes are mutually exclusive. The outcome can be only one or the other. Thus, using the division rule, the probability that any toss of the coin will produce a head is 1 (the outcome of interest) divided by 2 (the number of possible outcomes), or 1/2 (0.5). The probability that any toss will result in a tail is, of course, also 1/2 (0.5) and is deduced in the same manner as above. Or we may compute the probability of a tail in this manner: let $p$ = the probability of a head and $q$ = the probability of a tail. Since the probability that the outcome will be one or the other (a head or a tail) is 2/2 = 1.00, then

$p + q = 1$, and
$q = 1 - p$, or
$q = 1 - 0.5 = 0.5$

A **probability distribution** is a listing, in the form of a table or graph, of the probabilities associated with each possible outcome. A probability distribution that describes a single toss of the coin is:

| Outcome | Probability |
|---|---|
| head | 0.5 |
| tail | 0.5 |

A graph of this probability distribution is shown in figure 5.1.

**Figure 5.1** Probability distribution of the outcomes of the toss of a single coin

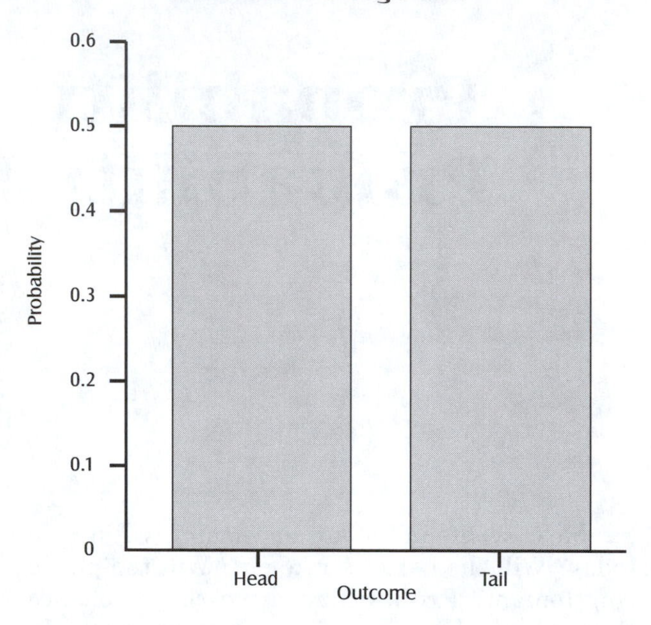


Note that we need not toss any coins to arrive at this probability distribution. It is a theoretical distribution that we derive from our knowledge of how flipped coins behave and from the laws of probability. Such a probability distribution is useful in many ways, not the least of which is to predict what we expect to happen in the real world. If we tossed a coin 100 times, we would expect to obtain approximately 50 heads and 50 tails if our assumptions about coin behavior under these conditions are correct. If we toss a die, the probability of obtaining a 1 is 1/6 (0.1666), the probability of obtaining a 2 is 1/6 (0.1666), and so on, for each of the 6 possible outcomes. The probability distribution for the outcomes of a single toss of a die is shown below in a table and in figure 5.2.

| Outcome [$x$] | 1 | 2 | 3 | 4 | 5 | 6 |
|---|---|---|---|---|---|---|
| Probability [$p(x)$] | 1/6 | 1/6 | 1/6 | 1/6 | 1/6 | 1/6 |

Again, note that we do not have to roll a lot of dice to arrive at this probability distribution. We deduce it from our knowledge of dice and the laws of probability. It is not based on observations of real events.

Next we'll consider repeating such experiments over more than one trial. We are now

**Figure 5.2** Probability distribution of the outcomes of the toss of a single die

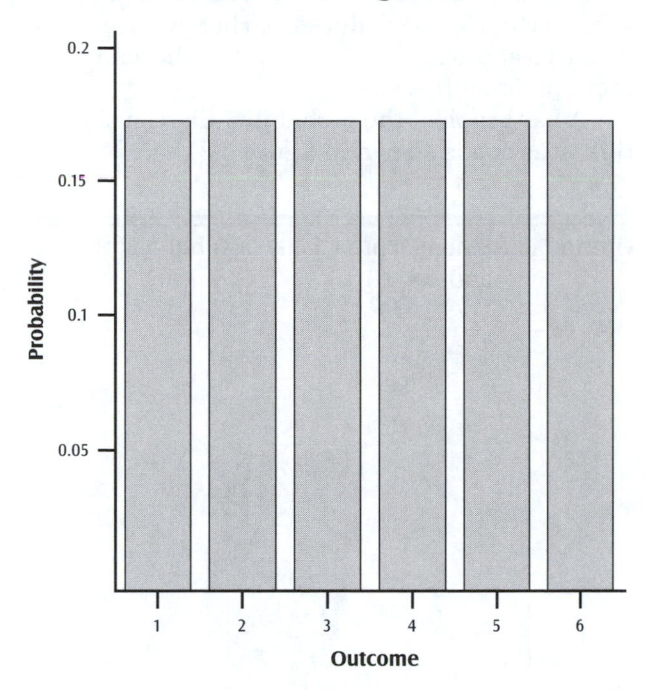


interested in the probabilities associated with the possible number of occurrences of the event in question. Several probability distributions describe the behavior of discrete variables. Two of these discrete probability distributions are the binomial distribution and the Poisson distribution and are considered in this chapter.

## 5.2 The Binomial Distribution

Many objects or events might be thought of as belonging to one of two mutually exclusive categories. Heads or tails, male or female, present or absent, sick or well, and asleep or awake are obvious examples. Even when there are more than two mutually exclusive categories, it is sometimes possible and useful to "lump" various categories into two by considering the object or event of interest as one category and everything else as the other category. Thus, "red" and "not red" are two mutually exclusive categories, even though "not red" can consist of blue, green, yellow, or any of a large number of colors that are not red. In the toss of a die, we might designate "1" as one possible outcome and "not 1" (any result except 1) as the other alternative.

The **binomial probability distribution** describes the probabilities associated with the outcome of certain events that have the following properties:

1. The event (the toss of a coin, for example), sometimes called a trial, occurs a specified number of times, designated as $k$.
2. Each time the event occurs, there are 2 mutually exclusive outcomes (a head or a tail). One of these outcomes is specified as the outcome of interest (sometimes arbitrarily called a "success"). The probability of this outcome is designated as $p$. The probability of the other outcome (a "failure" in statistical jargon) is designated as $q$. Since there are only 2 possible outcomes, $p + q = 1$ and $q = 1 - p$.
3. The events are independent, meaning that the outcome of one event has no influence on the outcome of other events in the series of events.
4. The number of times that the outcome of interest (a "success") occurs in $k$ events is designated as $x$. The binomial probability distribution gives the probabilities of $x$, symbolized by $p(x)$, for values of $x$ from zero to $k$.

Consider the possible events that can take place when you toss 2 coins simultaneously. The result can be 2 heads, 1 head and 1 tail, or 2 tails. What would the probability distribution for this situation look like? In other words, what is the probability of each of the 3 possible outcomes? The probability that one coin will be a head is 0.5 (1/2), and the probability that the other coin will be a head is also 0.5 (note that these 2 probabilities are independent of each other—the way that one coin lands has no effect on the way that the other coin lands). The probability that both will be heads is the product of their individual probabilities, which is $0.5 \times 0.5$, or 0.25.

**The Multiplication Rule (the "And" Rule)**

The probability that independent events will occur simultaneously (that event A and event B will both occur) is the product of the probabilities that the events will occur individually. Stated another way: if A and B are independent events with probabilities p(A) and p(B), the probability of the joint occurrence of both A and B is $p(A \text{ and } B) = p(A) \times p(B)$.

In the same manner, we determine the probability of 2 tails to be 0.25. However, there are 2 ways that the occurrence of one head and one tail can result. The first coin can be a tail ($p = 0.5$) and the second coin can be a head ($p = 0.5$), and the probability of this result is $0.5 \times 0.5$, or 0.25. On the other hand, the first coin can be a head ($p = 0.5$) and the second coin can be a tail ($p = 0.5$). The probability of this result is, as before, 0.25. Since the probability of a head and a tail is 0.25 and the probability of a tail and a head is 0.25, the probability of this event happening in either of these 2 ways is $0.25 + 0.25 = 0.5$.

### The Addition Rule (the "Or" Rule)

The probability that at least 1 of 2 or more alternative outcomes will occur is the sum of the individual probabilities that the events will occur. In other words, if A and B are mutually exclusive events with probabilities $p$(A) and $p$(B), the probability that either event A or event B will occur is $p(\text{A or B}) = p(\text{A}) + p(\text{B})$.

It might help to visualize this situation in a table, as shown below. The rows represent the possible outcomes of one coin, and the columns represent the possible outcomes of the other coin. We compute the probabilities of the possible joint occurrences of the 2 coins simply by multiplying the columns by the rows. (This approach may remind you of the Punnett square treatment in genetics. That is because it is the same concept!)

| | | The Other Coin | |
|---|---|---|---|
| | | H ($p = 0.5$) | T ($p = 0.5$) |
| One Coin | H ($p = 0.5$) | HH ($p = 0.5 \times 0.5$) | HT ($p = 0.5 \times 0.5$) |
| | T ($p = 0.5$) | TH ($p = 0.5 \times 0.5$) | TT ($p = 0.5 \times 0.5$) |

We may deduce these probabilities in an intuitive manner, as we did above, or we may simply expand the binomial expression:

$$(p + q)^k = 1 \qquad (5.1)$$

in which $p$ is the probability of one outcome of interest (heads), $q$ is the probability of the other outcome (tails), and $k$ is the number of events. In our example, $p$ is 0.5 (the probability that any coin toss will result in a head), $q$ is $1 - p$ ($1 - 0.5 = 0.5$), and $k$ is 2 (the number of coins tossed). Expanding the binomial expression above gives

$$p^2 + 2pq + q^2 = 1, \text{ or}$$

$$0.25 + 0.50 + 0.25 = 1$$

Each term in the expanded binomial expression gives the probability of one of the possible outcomes. Notice that these probabilities are the same as those we deduced earlier with a more intuitive approach. Notice also that the sum of all these probabilities is 1.

A bar graph of the probability distribution for this situation is shown in figure 5.3.

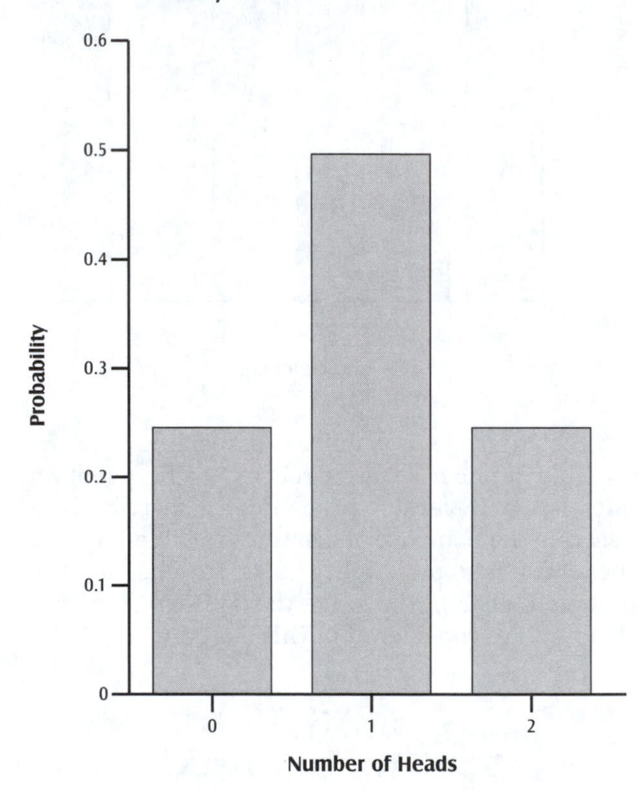


**Figure 5.3** Binomial probability distribution for $k = 2$ and $p = 0.5$

What are the probabilities of the various outcomes when 3 coins are tossed ($k = 3$)? The possible outcomes are: 3 heads, 2 heads and 1 tail, 1 head and 2 tails, and 3 tails. As before, $p$ (the probability that any coin will be a head) is 0.5, and $q$ is $1 - p$, or 0.5. Expanding the binomial expression gives

$$(p + q)^3 = 1$$

gives $$p^3 + 3p^2q + 3pq^2 + q^3 = 1, \text{ or}$$

$$0.125 + 0.375 + 0.375 + 0.125 = 1$$

These then are the probabilities associated with each of the 4 possible outcomes (0.125 for 3 heads, 0.375 for 2 heads and 1 tail, and so on). This probability distribution is shown as a bar graph in figure 5.4.

**Figure 5.4** Binomial probability distribution for $k = 3$ and $p = 0.5$

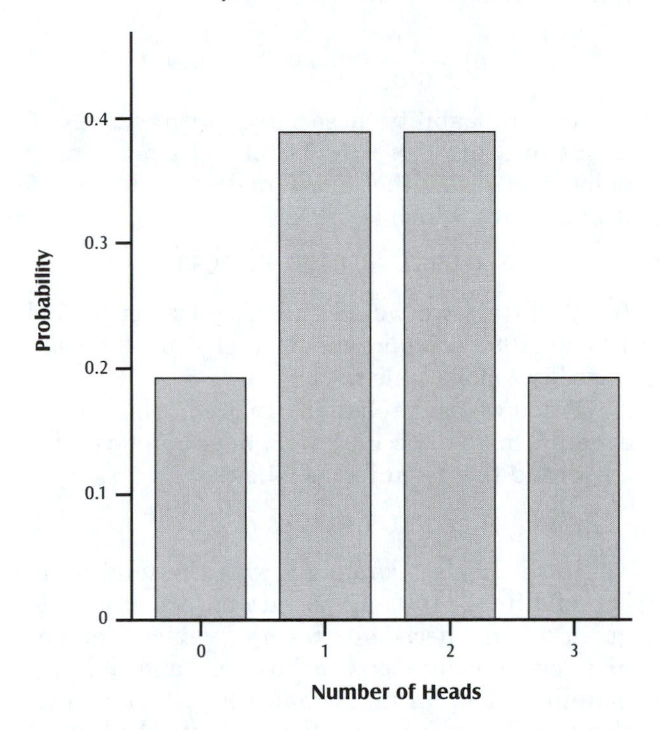


You may recall from a course in algebra that expanding binomial expressions is not a lot of fun, especially when $k$ becomes large. Fortunately there are better ways. We can use the binomial formula (below) to calculate exact probabilities. When $k$ gets really large (say > 30), we can also use the normal approximation for the binomial (section 6.4). The probability of the occurrence of $x$ events of interest (heads, for example), symbolized by $p(x)$, in $k$ trials (coins tossed, for example), is given by equation 5.2 (also called the **binomial formula**).

$$p(x) = \frac{k!}{x!(k-x)!} p^x q^{(k-x)} \qquad (5.2)$$

where $k$ is the number of trials (coins tossed), $x$ is the number of occurrences of the event of interest whose probability we wish to predict (the number of heads), and $p$ is the probability associated with the occurrence of $x$ (0.5 in this case). Formula 5.2 is derived from two of the rules introduced above—the multiplication rule (second part of the formula) and the addition rule (first part). Some of you may recognize the first part of the formula (also called the "binomial coefficient") as the number of combinations of $k$ objects taken $x$ at a time. Some examples will help to illustrate how equation 5.2 is used.

### • Example 5.1
Calculating a binomial probability

Suppose we wish to know the probability of obtaining 1 head in the toss of 3 coins (either 3 coins tossed simultaneously or 1 coin tossed 3 times in succession—in other words, in 3 trials). In this case $x = 1$ (the number of heads in which we have an interest), $k = 3$ (the number of trials), $p = 0.5$ (the probability associated with event $x$), and $q = 1 - p = 0.5$.

Substituting these values in equation 5.2 gives:

$$p(x=1) = \frac{3!}{1!(3-1)!}(0.5^1)(0.5^2) = 0.375$$

To verify that you understand the use of equation 5.2, find $p(x)$ for the other 3 possible outcomes in this situation (the probability of 0 heads and 3 tails is 0.125, of 1 head and 2 tails is 0.375, of 2 heads and 1 tail is 0.375, and of 3 heads and 0 tails is 0.125). Remember the use of factorials (e.g., $3! = 3 \times 2 \times 1 = 6$) and that, by definition, $0! = 1$.

So far we have considered only situations in which $p$ and $q$ are equal (both 0.5). This is not always the case. Consider the probability distribution of obtaining $x$ number of ones in the toss of 2 dice. We will consider a "1" to be one event of interest and "not 1" as the other event. The probability of obtaining any particular number on the roll of a single die is 1/6 (0.1667). Thus, $p$ (the probability of a 1) is 0.1667, and $q$ (the probability of obtaining any result except 1) is 1 – 0.1667, or 0.8333. Since we are tossing 2 dice in this experiment, $k = 2$. Let's again calculate one of the binomial probabilities, say for exactly 1 "ones." Substituting these values in equation 5.2 gives this result:

$$p(x=1) = \frac{2!}{1!(2-1)!}(0.1667^1)(0.8333^1) = 0.2778$$

The binomial distribution, showing the probabilities for all possibilities, is shown below. The sum of all these probabilities should equal 1, as it does.

| Number of "Ones" ($x$) | Number of "Not Ones" | $p(x)$ |
|---|---|---|
| 2 | 0 | 0.0278 |
| 1 | 1 | 0.2778 |
| 0 | 2 | 0.6944 |
| | Sum | 1.000 |

Thus, the probability of obtaining 2 ones in a toss of 2 dice is only 0.0278 (or approximately 3 times in 100 tosses). This probability distribution is shown as a bar graph in figure 5.5.

**Figure 5.5** Binomial probability distribution for $k = 2$ and $p = 0.1667$

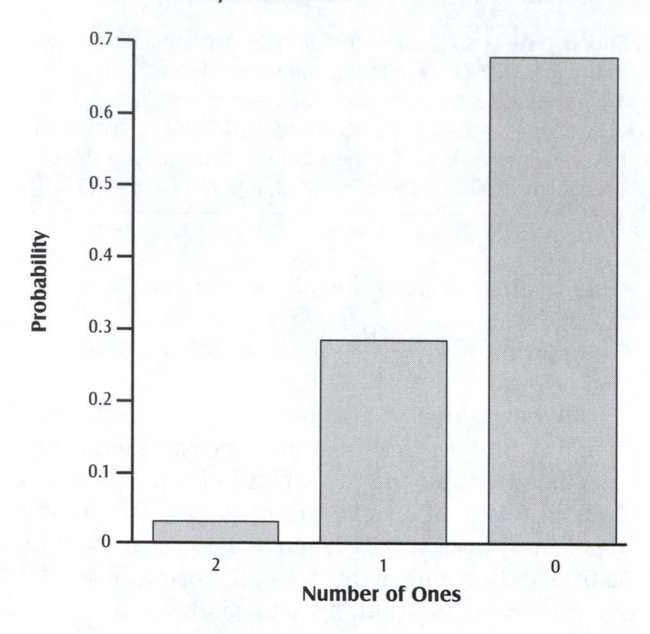


Compare this with figure 5.3, the probability distribution for $k = 2$ and both $p$ and $q = 0.5$. Notice the difference in shape of the distribution. Why is one symmetric and the other skewed?

Of what practical use is the binomial distribution to us as biologists? Perhaps some examples will serve to illustrate some of the many ways we use this particular probability distribution.

## • Example 5.2

### A binomial variable

Bush hogs always have litters of 6 hoglets. We will assume, based on our theoretical knowledge of the genetics of chromosomal sex determination, that the probability that any hoglet in a litter will be a male is 0.5 and the probability that any hoglet will be a female is also 0.5. We wish to know how many litters in 100 litters will consist of 5 or more males, if our assumption about the equal probability of an individual hoglet being a male or a female is correct.

Note that this problem is exactly equivalent to asking how often we would obtain 5 or more heads in the toss of 6 coins. First we solve equation 5.2 for $x = 5$, which is:

$$p(5) = \frac{6!}{5!1!}(0.5^5)(0.5^1) = 0.0937$$

Thus, the probability associated with exactly 5 males in a litter of 6 is 0.0937. However, our question asked for the probability associated with 5 *or more* males in a litter of 6. So we must also determine the probability associated with 6 males in a litter, which is:

$$p(6) = \frac{6!}{6!0!}(0.5^6)(0.5^0) = 0.0156$$

The probability associated with exactly 6 males in a litter of 6 is 0.0156. The probability associated with either 5 *or* 6 males in a litter of 6 is $p(5) + p(6)$, which is

$$0.0937 + 0.0156 = 0.1093$$

In 100 litters we would therefore expect to find $100 \times 0.1093$ or approximately 11 litters consisting of 5 or more male hoglets.

When we use probability to predict how many events might occur, we have determined **expected frequencies** as follows:

$$E(x) = p(x) \times n \tag{5.3}$$

In the hoglet example, $n$ was the total number of litters (100). Expectations are used frequently in decision theory, which guides management decisions in business and military planning. Keep in mind that, because of chance, these decisions are not always right! (And when we criticize these decisions after the fact, we are guilty of "armchair quarterbacking.")

## • Example 5.3

### Example 5.2 viewed in a different context

Let us ask the same question in a slightly different way: In 100 litters, how many of each of the possible combinations of males and females would we expect to find (how many with 6 males and no females, with 5 males and 1 female, with 4 males and 2 females, and so on)?

For this solution we would simply solve equation 5.2 for each value of $x$ from 6 to 0. We would obtain the results shown in table 5.1. Notice that the sum of the $p(x)$'s is close to 1. (The slight discrepancy is from a rounding error.)

What use might we wish to make of such a probability distribution, which is based on theoretical assumptions rather than on real-world observations? Consider example 5.2. Suppose we suspect that the gender of bush hoglets is not a matter of random chance but rather that some females tend to produce an inordinately large number of offspring of one sex or the other (a phenomenon that has actually been observed in a number of species). Suppose also that we actually surveyed 100 litters of bush hogs, each consisting

**Table 5.1** Binomial distribution for example 5.3

| Number of Males ($x$) | Number of Females ($k-x$) | $p(x)$ | Expected Frequency $p(x) \times 100$ |
|---|---|---|---|
| 6 | 0 | 0.0156 | 1.56 |
| 5 | 1 | 0.0937 | 9.37 |
| 4 | 2 | 0.2344 | 23.44 |
| 3 | 3 | 0.3125 | 31.25 |
| 2 | 4 | 0.2344 | 23.44 |
| 1 | 5 | 0.0937 | 9.37 |
| 0 | 6 | 0.0156 | 1.56 |
| | **Sum** | 0.9999 | 99.99 |

of 6 hoglets, and found that 20 or so contained 5 or 6 male hoglets and a similar number contained 5 or 6 female hoglets. This is about twice as many as our binomial probability distribution predicts, and therefore, we would have reason to suspect that sex determination in this species is not a random event. In other words, some factor or factors other than the chance combination of X and Y chromosomes is involved.

Note carefully that without the prediction based on the binomial probability distribution, we would not know if the number of predominately single-sex litters that we observed was too large to be produced by random combinations of X and Y chromosomes. In other words, we cannot conclude that a result is *not* due to chance alone unless we know what result is expected if it is due to chance. This concept is the basis for much of statistical hypothesis testing, the subject of chapter 8.

## 5.3 The Poisson Distribution

Another discrete probability distribution often useful to biologists is the **Poisson distribution**. Generally we use the Poisson distribution to predict probabilities of the occurrences of "rare" events (when such occurrences are known to be independent of one another) or to determine if the occurrences of such events are independent of one another. The event of interest must be rare, which is to say that its occurrence in any sampling unit must be small *relative to the number of times that it could occur.* For an event to follow a Poisson distribution, occurrences of the event must be independent of previous occurrences of the event in the sampling unit. We can compare an observed frequency distribution with frequencies produced by a Poisson distribution to see if occurrences are independent. These occurrences may be in space (e.g., locations of plants) or in time (e.g., when a bird sings).

### • Example 5.4

A Poisson variable: spatial distribution of plants

In the maple seedling example in chapter 3 (reproduced below as table 5.2), most of the quadrats sampled had no seedlings, several had 1, a few had 2, and so on. We would like to know if the occurrences of maple seedlings are independent events. If so, their frequency of occurrence in the quadrats should follow a Poisson distribution. If they do not follow a Poisson distribution, we conclude that the events are not independent and that the occurrence of a maple seedling in a quadrat in some way influences the occurrences of other maple seedlings in that quadrat.

**Table 5.2** Maple seedlings per quadrat

| Number of Plants/Quadrat | Frequency |
|---|---|
| 0 | 35 |
| 1 | 28 |
| 2 | 15 |
| 3 | 10 |
| 4 | 7 |
| 5 | 5 |

Since the number of seedlings that could conceivably occur in a quadrat is quite large, we can consider the observed frequencies to be rare events. If the presence of a seedling in a quadrat does not alter the possible occurrence of other seedlings in a quadrat, the frequency of seedlings per quadrat should follow a Poisson distribution. If the observed frequency distribution strongly deviates from the Poisson distribution, then we have reason to suspect that the occurrences of maple seedlings are not independent events, which is to say that the distribution of maple seedlings is not random, and that some factor or factors in the environment other than chance influences the distribution.

What the Poisson distribution predicts is how many units (quadrats, in this case) are expected to have no occurrences of the event (maple seedlings), how many are expected to have 1, how many are expected to have 2, and so on, if the events are rare and independent. These probabilities are given by the expression:

$$p(x) = \frac{\mu^x e^{-\mu}}{x!} \tag{5.4}$$

where is the population mean occurrence of the event per sampling unit, $e$ is the base of natural logarithms (2.7183), and $p(x)$ is the probability of $x$ events in a unit. Since we usually have no knowledge of the population mean, it is estimated by the sample mean, and $p(x)$ is computed by equation 5.5.

$$p(x) = \frac{\bar{x}^x e^{-\bar{x}}}{x!} \tag{5.5}$$

for probabilities, $p(x)$, of 0, 1, 2, 3, 4, and so on, occurrences per unit. In the maple seedling example, there were 100 quadrats (units) and a total of 141 seedlings (events) in these 100 quadrats. The mean number of occurrences ($\bar{x}$) was therefore 141/100 or 1.41 plants per quadrat. Solving equation 5.5 for relative expected occurrences per unit (seedlings per quadrat) of 0, 1, 2, 3, 4, 5, and 6, gives the results shown in table 5.3 under the column designated "$p(x)$." For example, one of these probabilities (1 seedling/quadrat) was determined as follows:

$$p(x=1) = \frac{(1.41^1)(e^{-1.41})}{1!} = 0.344$$

Multiplying these values by the number of sampled units (100 in this case) gives the expected frequencies (from formula 5.3). The last column in the table gives the values that were actually observed (from table 5.2). The column headed "Cumulative" is the cumulative probability for the designated value of $x$ or any smaller value (the sum of $p(x)$ for the current row and all rows above it). It is occasionally convenient to make use of this cumulative distribution.

**Table 5.3** Expected and observed frequencies of maple seedlings per quadrat ($\bar{x}$ = 1.41, $n$ = 100)

| Number/ Quadrat ($x$) | $p(x)$ | Cumulative | Expected | Observed |
|---|---|---|---|---|
| 0 | 0.244 | 0.244 | 24.4 | 35 |
| 1 | 0.344 | 0.588 | 34.4 | 28 |
| 2 | 0.243 | 0.831 | 24.3 | 15 |
| 3 | 0.114 | 0.945 | 11.4 | 10 |
| 4 | 0.04 | 0.985 | 4 | 7 |
| 5 | 0.011 | 0.997 | 1.1 | 5 |
| 6 | 0.003 | 0.999 | 0.3 | 0 |

Figure 5.6 is a graphic representation of the expected distribution and the actual observed distribution from table 5.3. The expected distribution, based on the Poisson probability distribution, is the distribution we would expect if the maple seedlings were distributed randomly within the sampled habitat.

**Figure 5.6** Comparison of observed frequencies of maple seedlings per quadrat (from table 5.2) with frequencies expected from a Poisson probability distribution for mean = 1.41

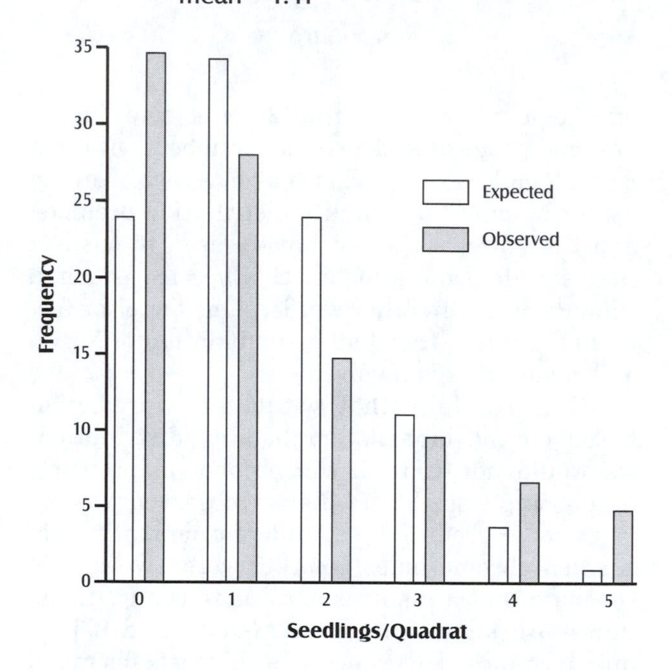


A frequent use of the Poisson distribution is testing whether particular events occur independently of one another, and therefore are distributed randomly in space or time. When the events are not random, they may either be "clumped" or "repulsed" with respect to each other. Notice in figure 5.6 that the expected (Poisson) distribution and the actual observed distribution do not match very well. We might suspect, therefore, that the occurrences of maple seedlings are not independent of one another. In chapter 14 we will discuss how a goodness-of-fit test can be applied to determine if this difference is likely real or if it is simply the result of chance. This example illustrates one way in which the Poisson distribution is commonly used in ecology, but this distribution also finds use in many other situations. Example 5.5 illustrates how the Poisson distribution might be applied to a microbial genetics problem.

### • Example 5.5

A Poisson variable: Viruses infecting bacteria

Bacterial viruses infect bacterial cells by first adsorbing onto the bacterial cell wall. The number of phage particles that adsorb to any one bacterial cell is a Poisson process. In a certain experiment, $2.5 \times 10^6$ phage particles were mixed with $10^6$ bacterial cells. Thus, the mean number of phage particles per bacterial cell is 2.5. We wish to know what proportion of the bacterial culture would be expected to have no phage particles adsorbed to them.

The mean number of phage particles per bacterial cell is 2.5. Substituting this value in equation 5.5 and solving for $p(0)$ gives:

$$p(0) = \frac{2.5^0 \times e^{-2.5}}{0!} = 0.0821$$

Thus, the probability that any randomly selected cell would have no phage particles adsorbed to it is 0.0821. This is also the proportion of the cells in the bacterial culture that would be expected to be free of phage particles. The expected number of bacterial cells with no phage particles should therefore be $0.0821 \times 10^6 = 8.21 \times 10^4$.

A question like this can be asked in a different way. Suppose we wish to add a sufficient number of phage particles to a bacterial culture to ensure that not more than 1% (0.01) of the bacterial cells remain uninfected. How many phage particles per bacterial cell would we need to add? Could you suggest how this might be calculated?

## Key Terms

binomial distribution
cumulative distribution
expected frequency
Poisson distribution
probability
probability distribution

## Exercises

5.1 In the toss of 2 coins, what is the probability of obtaining 2 heads? (Since there are only two possibilities, notice that this question could also be asked: "What is the probability of obtaining 0 tails?".)

5.2 In the toss of 2 coins, what is the probability of obtaining 2 tails?

5.3 In the toss of 2 coins, what is the probability of obtaining 1 head and 1 tail?

5.4 In the toss of 6 coins, what is the probability of obtaining 2 heads and 4 tails?

5.5 In the toss of 6 coins, what is the probability of obtaining 2 or fewer tails?

5.6 In the toss of 5 dice, what is the probability of obtaining 3 twos?

5.7 In the toss of 7 dice, what is the probability of obtaining at least 2 ones?

5.8 In the toss of 7 dice, what is the probability of obtaining not more than 2 ones?

5.9 Assume that the probability that any child born in a population will be a boy is 0.5 and the probability that it will be a girl is also 0.5. For a randomly selected family of 6 children, what is the probability that there will be exactly 3 boys and 3 girls?

5.10 For the situation in exercise 5.9, what is the probability that there will be 2 or fewer girls?

5.11 For the situation in exercise 5.9, what is the probability that there will be exactly 6 boys?

5.12 For the situation in exercise 5.9, what is the probability that there will be exactly 6 girls?

5.13 In couples where each person is heterozygous for the sickle-cell gene, there is a probability of 0.25 that any child of the couple will actually have the disease, and a probability of 0.75 that any child will not have the disease. For a population of families of 6 children in which both parents are heterozygotes, what is the probability that in a randomly selected family, none of the 6 children will have the disease?

5.14 Construct a table and a bar graph of the probabilities of all possible outcomes of the situation in exercise 5.13 (no children have the disease, 1 child has the disease, 2 children have the disease, and so on).

5.15 For the situation in exercise 5.13, compute the probability that 1 or more children will have the disease.

5.16 For the situation in exercise 5.13, compute the probability that 2 or more children will have the disease.

5.17 The proportion of male births to female births in a population is 49:51. Thus, the probability that any individual born into this population will be a male is 0.49 (49/100). Suppose families consist of 8 offspring. Determine the binomial distribution for this example. First do by hand. Next, if you have a statistical package available, use the computer to generate this distribution. (In MINITAB, choose calc, then probability distributions; you'll need to have the integers 0, 1, 2, . . . , 8 in Cl.)

5.18 For the situation in exercise 5.17, compute the expected frequency of families (out of 1000) in which there are exactly 4 male and 4 female offspring.

5.19 For the situation in 5.17, compute the expected frequency of 1000 families in which all 8 offspring are male.

5.20 Suppose it is known that in a certain population 10% of the population is colorblind. If a random sample of 25 people is drawn, find the probability that:

a) none will be colorblind

b) one or more will be colorblind

5.21 Using the data from exercise 2.3 (see chapter 2), compute the Poisson probabilities for $x = 0$ to $x = 8$ for a population whose mean is the same as this sample mean. Calculate the expected frequencies and compare to the observed frequencies in a table. Save your results for later use.

5.22 Construct a bar graph of the expected frequencies and observed frequencies for these results from exercise 5.21. Save your results for later use.

5.23 Interpret the data from exercise 5.21. Find the probability that a quadrat sampled at random in this population will contain exactly:

a) 8 ant lion pits

b) 5 or more ant lion pits

5.24 In a culture containing $3 \times 10^7$ bacterial cells and $5 \times 10^7$ bacteriophages, what proportion of the bacterial cells would be expected to have:

a) no phage particles adsorbed to their cell walls?

b) 1 or more phage particles adsorbed to their cell walls?

5.25 In a culture containing $5 \times 10^8$ bacterial cells and $3 \times 10^8$ bacteriophages, what proportion of the bacterial cells would have 3 or more phage particles adsorbed to their cell walls?

# 6

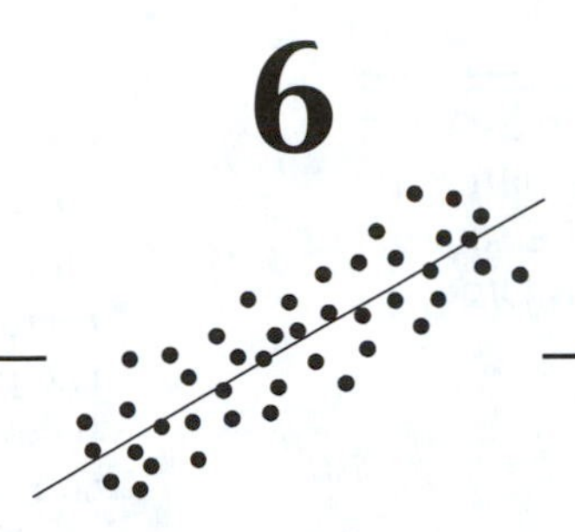

# The Normal Distribution

The **normal distribution** is a continuous probability distribution that is widely used in statistics. It closely approximates a wide variety of continuous measurement variables, such as heights, weights, and agricultural yields. Many discrete variables also approximately follow such a distribution when the range of values that the variable assumes is fairly large (see "Normal Approximation of the Binomial Distribution", section 6.4). We can also use the normal distribution to easily determine probabilities. Thus, the normal distribution underlies a number of important statistical procedures, including t-tests, ANOVA, and regression analysis. In this chapter, we explore some properties and uses of the normal distribution.

## 6.1 The Normal Distribution and Its Properties

Continuous measurement variables can assume any value between certain limits. For example, recall the mosquito-fish lengths in table 3.4. An individual fish 25 mm long could, if measured with a fine enough ruler, actually be 25.01253274 mm. The histogram we generated from 172 measurements (figure 3.3) was informative, but somewhat crude. Imagine if we instead had collected and measured a very large number of fish (say a million). If we then divided these measurements into many bins, each very thin, the histogram would become a smooth curve, somewhat like figure 6.1. The probability distribution that describes curves of this general type is the normal probability distribution.

**Figure 6.1** A normal distribution, centered at the population mean, $\mu$. Since the normal curve extends to $\pm\infty$, the curve approaches the x-axis as an asymptote

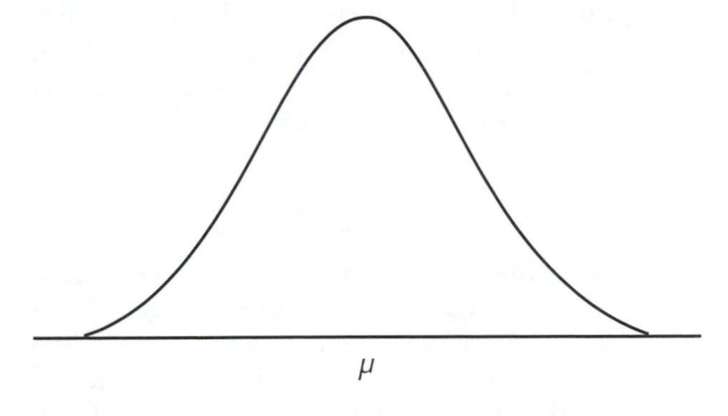


The normal probability distribution has a number of important characteristics, which are discussed below. Reference to figures 6.2 and 6.3 will assist in understanding these characteristics.

**Figure 6.2** Three normal distributions. Curves A and B have the same variance, but different means; curves B and C have the same means, but the variance of B is larger than that of C

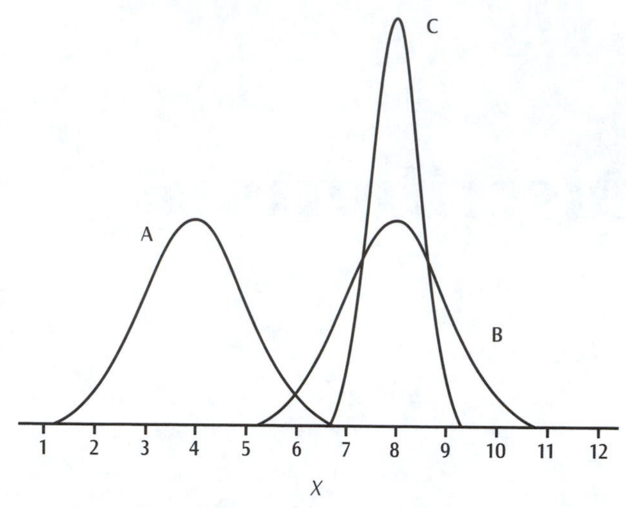


From Sokal and Rohlf 1995

### *Properties of the Normal Distribution*

1. The distribution is completely defined by the mean ($\mu$) and the standard deviation ($\sigma$). The location of the curve along the $x$-axis is defined by the mean, and the spread of the curve is defined by the standard deviation (figure 6.2). Since these parameters can assume, theoretically, an infinity of values, there are an infinite number of normal distributions.
2. The height on the $y$-axis of any point along the curve ($x$) represents the probability density function, $f(x)$, of that value of the variable. The probability density function is expressed formally as:

$$f(x) = \frac{1}{\sigma\sqrt{2\pi}} e^{-1/2(x-\mu/\sigma)^2} \quad (6.1)$$

We cannot think of this $y$-axis value as representing a probability (or frequency) since a point has no dimension, and therefore, a line drawn vertically from such a point has no width. Thus, there is no real value that corresponds to this value of $x$. However, if we consider two values of $x$ that are *very* close together, we could compute the area of the curve that lies between these two values (using integral calculus). In this way we may assign a probability to a very narrow *range* (class interval) of the variable in question.

3. The curve is perfectly symmetrical about the mean. Thus, the mean and median are equal.
4. One standard deviation above the mean includes 34.13% of all of the individuals in the population, and one standard deviation below the mean includes 34.13% of all of the individuals in the population. Thus, 68.26% of all of the individuals in the population fall within plus or minus one standard deviation of the mean (figure 6.3). Another way of thinking about this is that there is a probability of 0.6826 that any individual ($x$) taken at random from this population would fall within one standard deviation of the mean. In a similar manner, the mean ±2 standard deviations contains 95.46% of all of the individuals in the population (or, the probability that any individual taken at random from this population would fall within ±2 standard deviations of the mean is 0.9546). The mean ±3 standard deviations contains 99.73% of all the individuals within the population (table 6.1). We will have occasion to examine this property in considerably more detail in this and subsequent chapters.

**Figure 6.3** Areas of a normal distribution

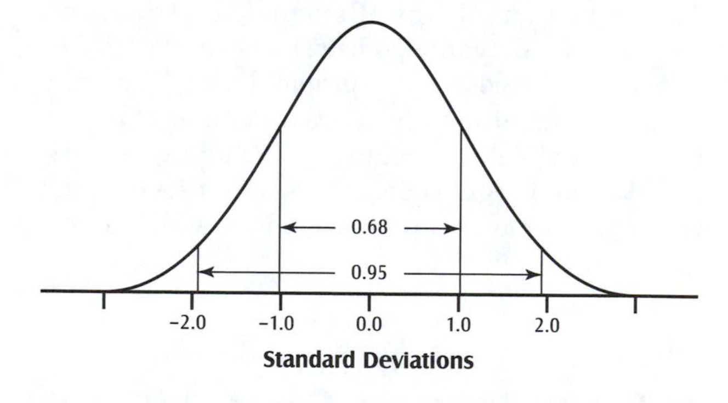


**Table 6.1** Proportions of the area under the normal curve as a function of the standard deviation

| Standard Deviations | Proportion of Area |
|---|---|
| ±1 | 0.6826 |
| ±1.960 | 0.9500 |
| ±2 | 0.9546 |
| ±2.576 | 0.9900 |
| ±3 | 0.9973 |

The total area under the normal curve is, by definition, exactly 1. Since the normal curve is defined by the mean and the standard deviation, there is a fixed relationship between the standard deviation and the proportion of the area under the curve that occurs between the mean and any standard deviation. The relationships shown in table 6.1 hold true for any normal distribution. We will later use an expansion of this table to find areas more exactly.

## 6.2 The Standard Normal Distribution and Z Scores

Fortunately, we do not have to deal with complex computations with the normal probability density function every time we wish to use a normal distribution. Among the infinite variety of normal distributions that are possible, there is one to which all others may be rather easily converted: the **standard normal distribution**. It has a mean of zero and a standard deviation of one, and its "tails" extend from negative infinity through zero (the mean) to positive infinity. Table A.1 in the appendix gives the proportion of the standard normal curve that lies between zero and almost any value of a variable that one might reasonably expect to encounter. We discuss how this table is used in the examples that follow.

We often need to determine where a particular value falls on a normal curve or, more often, the probability that some value or range of values lies within a certain portion of the curve. To do this, individual variants need to be expressed as if they were variants of the standard normal distribution. In other words, we need to convert our normal distribution, which may have any imaginable mean and standard deviation, to the standard normal distribution, which has a mean of zero and a standard deviation of one. In practice it is usually unnecessary to convert all of our observations; only those of interest need be converted. These converted observations are called ***z* scores**, standard scores, or standard normal deviants. Equation 6.2 transforms any variable into its corresponding $z$ score.

$$z = \frac{x - \mu}{\sigma} \tag{6.2}$$

What might not be immediately apparent here is that a $z$ score defines how far from the mean, in terms of standard deviations, an observation lies. An observation with a $z$ score of +1 is one standard deviation greater than the mean, and an observation with a $z$ score of −1 is one standard deviation less than the mean (figure 6.3). An observation with a $z$ score of 0 has the same value as the mean. Thus, any variant is expressed not in its actual value but in how far, in terms of standard deviations, it lies from the mean.

---

- **Example 6.1**

An approximately normally distributed variable

The height (in centimeters) of a large group ($n$ = 414) of female general biology students was determined. This group is the entire population in which we have an interest. The mean height of this group was 166.8 cm, and the standard deviation was 6.4 cm.

---

Height in humans is an approximately normally distributed variable, so we can use the standard normal distribution to answer several questions about this population.

***Question 1***

What is the $z$ score of an individual who is 170 cm tall?

$$z = \frac{x - \mu}{\sigma} = \frac{170 - 166.8}{6.4} = 0.5$$

It is easy enough to comprehend that this person is 3.2 cm taller than the average student in this population, so why is it of interest to us that she is 0.5 standard deviations taller than the average? Let us ask another question.

***Question 2***

What proportion of the population is as short as or shorter than this 170 cm tall person? A somewhat different way of asking this same question is: What is the probability that an individual sampled at random from this population will be 170 cm or shorter? Keep in mind in this and subsequent discussions that the area of a probability distribution corresponding to a designated proportion of the distribution may also be thought of as a probability. To solve this problem, we need to become acquainted with table A.1. When considering the use of table A.1 in this example, it will help to refer to figure 6.4.

Zero on the $x$-axis of this graph represents a $z$ score of 0, the population mean. The shaded area of the curve (the area between the mean and our $z$ score) represents the proportion of the standard normal distribution that falls between the mean and the indicated value of $z$. We can find that number on table A.1. The left column of the table

**Figure 6.4** The standard normal distribution (example 6.1, question 2)

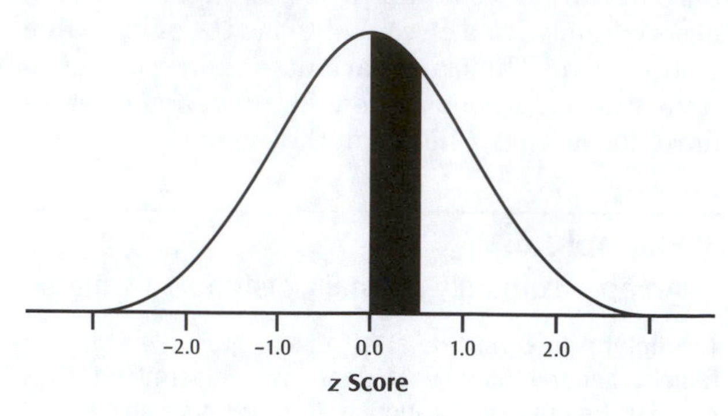


gives $z$ scores to 1 significant digit. The top row extends this to 2 significant digits. For example, if we wish to know the proportion of the curve that lies between the mean and a $z$ score of 1.23, we look down the left column to $z$ = 1.2 and go across to the column headed .03. The value we find there is 0.3907. Look this up yourself in appendix A right now.

Let us rephrase the original question: What proportion of the standard normal distribution has a $z$ score of 0.5 or less? We first need to determine what proportion of the standard normal distribution lies between the mean and a z score of 0.5 (figure 6.4). According to table A.1, the area is 0.1915 (verify this for yourself). Therefore, 0.1915 (or 19%) of the individuals in this population lie between 170 cm (our person with a $z$ score of 0.5) and 166.8 cm, which is the mean. Note that the standard normal distribution is perfectly symmetrical and that 0.5000 of the area lies above the mean and 0.5000 of the area lies below the mean. Our original question was: what proportion of the population is as short as or shorter than 170 cm? We know that 0.5000 (50%) lies below the mean (by definition). We also know that 0.1915 (19%) lies between the mean and a $z$ score of 0.5 (corresponding to 170 cm). Therefore, 0.5000 + 0.1915, or 0.6915 (roughly 69%), lies below a $z$ score of 0.5. We conclude that the proportion of the individuals in this population who are 170 cm or shorter is 0.6915, or in other words, approximately 69% of the individuals in this population are as short as or shorter than this individual who is 170 cm tall. This may also be interpreted to mean that the probability that any individual sampled from this population will be 170 cm or shorter is 0.6915.

You may be curious about why the 0.5000 of the standard normal curve that lies between the mean and negative infinity is not already added to the entries in table A.1. Another question might help to clarify this.

***Question 3***

What proportion of this population is expected to be 163.6 cm or shorter? The $z$ score for this observation is:

$$z = \frac{163.6 - 166.8}{6.4} = -0.5$$

Graphically, the problem looks like figure 6.5. We are trying to determine how much (what proportion) of the standard normal distribution lies to the left of (below) our $z$ score of –0.5. Since the standard normal distribution is symmetrical, the same proportion lies between the mean and a $z$ score of –0.5 that lies between the mean and a $z$ score of +0.5, which we already know is 0.1915. (If you do not remember how we already know this, refer back to the last question.)

**Figure 6.5** The standard normal distribution (example 6.1, question 3)

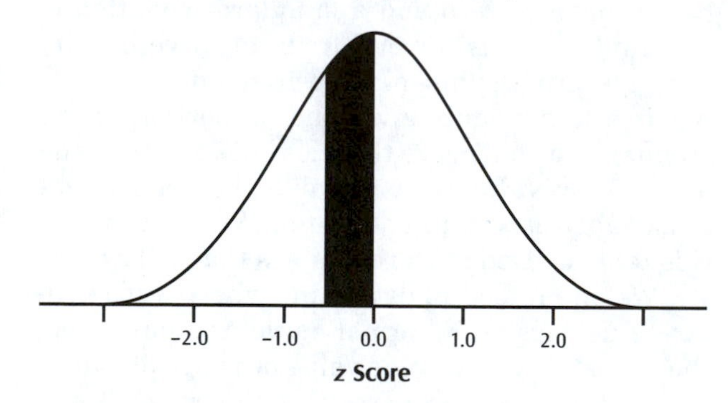


Thus, the proportion that lies below a $z$ score of –0.5 is:

$$0.5000 - 0.1915 = 0.3085$$

We conclude that approximately 31% of this population will be as short as or shorter than 163.6 cm, or that the probability that any individual sampled at random from the population will be 163.6 cm or shorter is 0.3085.

***Question 4***

What proportion of the population lies between 160 cm and 170 cm? This problem is shown graphically in figure 6.6. To solve it we first determine what proportion of the curve lies between the mean and 160 cm, which is:

$$z = \frac{160 - 166.8}{6.4} = -1.063$$

**Figure 6.6** The standard normal distribution (example 6.1, question 4)

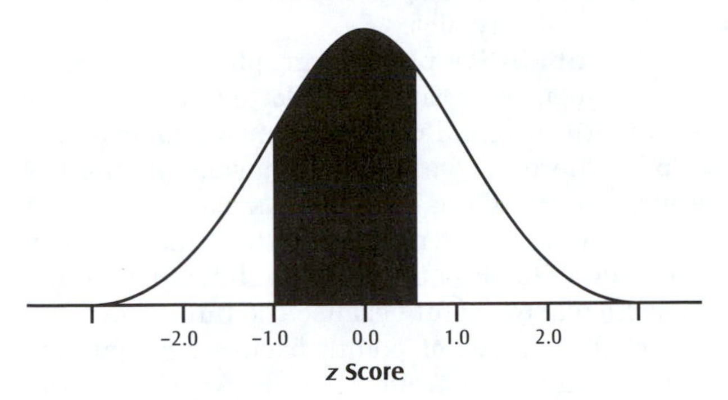


According to table A.1, 0.3554 of the standard normal curve lies between the mean and a $z$ score of –1.063 (–1.06, actually, since the table only goes to 2 places for $z$ scores). Next we determine how much of the curve lies between the mean and 170 cm, for which the $z$ score is:

$$z = \frac{170 - 166.8}{6.4} = 0.50$$

Again, according to table A.1, 0.1915 of the standard normal distribution lies between the mean and a $z$ score of 0.50. Thus, the proportion lying between 160 cm and 170 cm, or between a $z$ score of –1.06 and 0.50, is 0.3554 + 0.1915, or 0.5469. Approximately 55% of this population is expected to consist of individuals whose height is between 160 cm and 170 cm. The probability that any individual sampled at random from this population will be between 160 cm and 170 cm is 0.5469.

***Question 5***

What range of heights includes 0.95 (95%) of this population? This problem is somewhat different from the previous ones. Here we know the proportion of the standard normal distribution in which we have an interest (that is, we know what the $z$ score is, and we must find the value of $x$ that corresponds to this $z$ score). Table A.1 shows that 0.475 of the standard normal distribution lies between the mean and a $z$ score of 1.96, and therefore, 0.475 lies between the mean and a $z$ score of –1.96. Thus, 0.95 is between the mean and a $z$ score of ±1.96. (Refer to figure 6.3 to visualize this.) Since

$$z = \frac{x - \mu}{\sigma}$$

it follows that

$$x = z\sigma + \mu$$

Accordingly, the height ($x$) that includes 0.475 of the standard normal distribution below the mean is

$$x = (-1.96 \times 6.4) + 166.8 = 154.23$$

and the height ($x$) that includes 0.475 of the standard normal distribution above the mean is

$$x = (1.96 \times 6.4) + 166.8 = 179.34$$

We conclude that 95% of the individuals in this population are between the heights of 154.23 cm and 179.34 cm. Or, thinking of probabilities, there is a probability of 0.95 that any individual sampled at random from this population will be between 154.23 cm and 179.34 cm, and a probability of 0.05 that any individual sampled at random from this population will be taller than 179.34 cm or shorter than 154.23 cm.

Since we rarely know the population mean and standard deviation of a population, how useful the information that we have been considering? From an applied standpoint, not much. However, an understanding of—or at least an acquaintance with—these concepts gives one the proper frame of mind to learn about another group of probability distributions, called sampling distributions, and the underlying concepts of statistical inference.

## 6.3 Testing for Normality

Many of the statistical tests that we consider in the following chapters are based on the assumption that the variable in question is at least approximately normally distributed. Fortunately, the normal distribution describes, or at least approximates, a large number of biological variables. However, one should never assume that a variable is approximately normally distributed unless there is some reason to believe that this is the case. Example 6.1 asserted that height in humans is an approximately normally distributed variable. How do we know this? One way is to examine the frequency distribution of the variable from a large sample.

Consider the male heights given in exercise 3.1. The histogram for these data, using a class interval of 5 cm, is shown in figure 6.7. Note that, in general, this histogram is more or less bell-shaped, and its overall appearance is much like that of a normal distribution. For many applications this would provide sufficient justification for assuming that the distribution is approxi-

**Figure 6.7** Frequency distribution of human male heights (data from exercise 3.1)

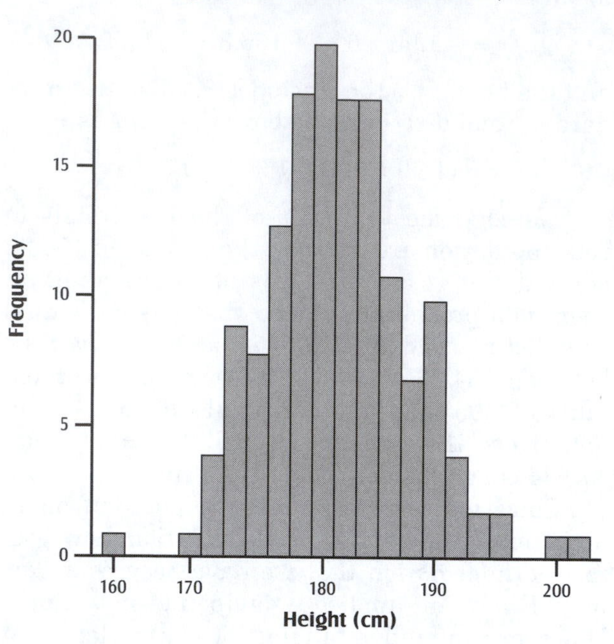


Data from exercise 3.1

mately normal, particularly if we were using a large sample (see chapter 7 to find out why). However, a better way to check for normality is to use a probability plot.

A **probability plot** is a graph of the cumulative frequency (section 3.3) plotted on the $y$-axis against the original measurement variable on the $x$-axis, where the $y$-axis uses a normal probability scale. The $y$-axis is scaled in such a way that when data are drawn from a normal distribution, the points fall close to a straight line. Such a plot is particularly useful because the human eye can detect departures of points from a straight line more easily than from a curve. An example of such a plot is shown in figure 6.8. Notice that the plotted points are quite close to the straight line, suggesting that this sample of human heights was taken from a population having a normal distribution. Doing these plots by hand is somewhat tedious. Fortunately, computers never get bored doing repetitive tasks. Many statistical packages contain programs for normality tests and probability plots. Besides showing the plot, some of these programs also perform a formal statistical procedure (a type of goodness-of-fit test) to test

**Figure 6.8** Probability plot for human male heights, generated from MINITAB

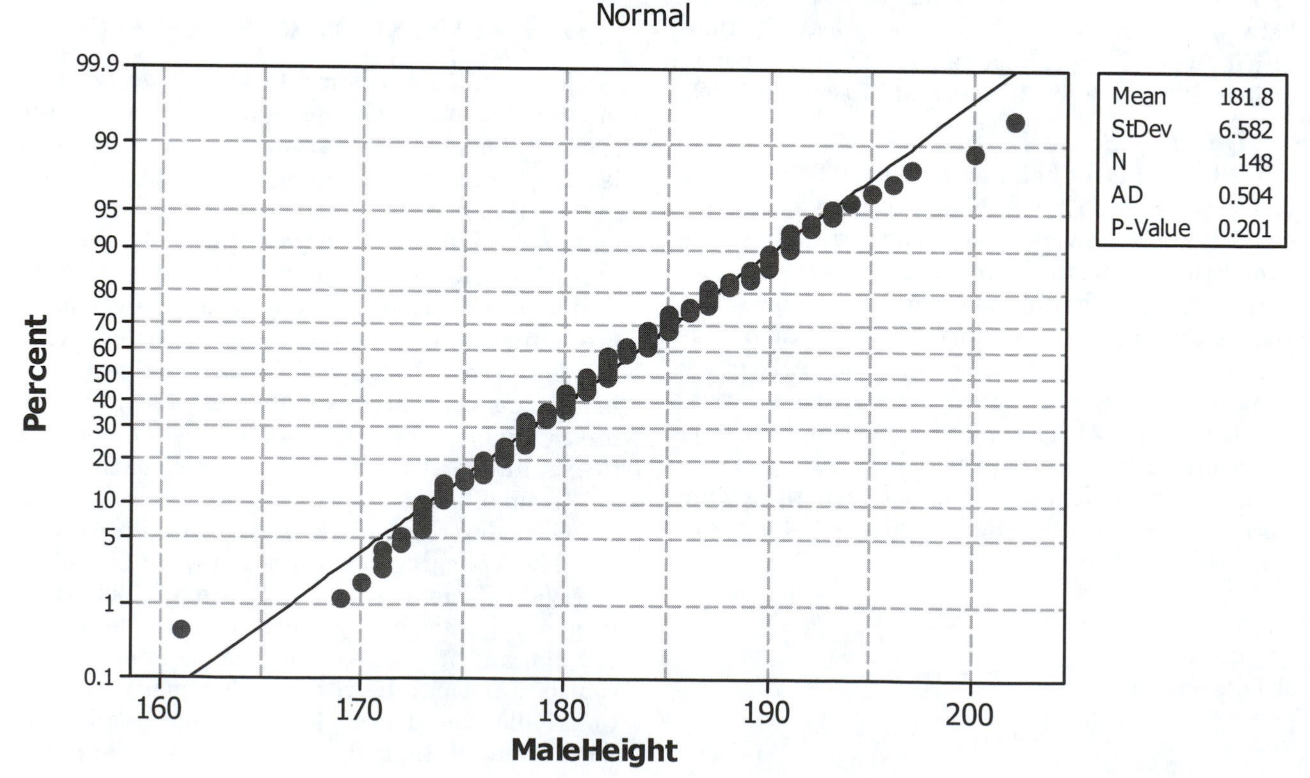

for normality. In figure 6.8, the Anderson Darling test (AD) results in a p-value of 0.201. In general, a large p-value (say > 0.05) for this test is consistent with a close fit of the data to a normal distribution. (We visit p-values in much more depth when we examine the nature of hypothesis tests in chapter 8.)

## 6.4 Normal Approximation of the Binomial Distribution

Refer back to the binomial distribution (section 5.2) and recall that we calculated the probability that $x$ has a specific value given that we know two parameters: $k$, the number of cases, and $p$, the probability of success in any one selection. We also know that $p + q = 1$. Recall also that these calculations can be somewhat tedious, particularly when $k$ is large. Fortunately, the normal distribution can sometimes be used to solve the problem more quickly.

When $k$ is fairly large and $p$ is not too close to zero or one, the binomial distribution becomes approximately a normal distribution. How large is "fairly large"? As a rule of thumb, if the products $k \times p$ and $k \times q$ are both equal to and larger than 5, the normal approximation is considered to be fairly accurate.

### • Example 6.2

Suppose that 25 individuals are sampled from a population in which the ratio of females to males is known to be 1:1, and we wish to know the probability of obtaining a sample of 8 or fewer males in this sample of 25 individuals. Let $p$ = the probability that any individual sampled will be a male (0.5), let $q$ = the probability that any individual sampled will be a female (0.5), and let $x$ = the number of males in which we have an interest (8 or fewer, in this case). The entire probability distribution is illustrated in figure 6.9.

**Figure 6.9** Binomial probability distribution for $k = 25$ and $p = 0.5$

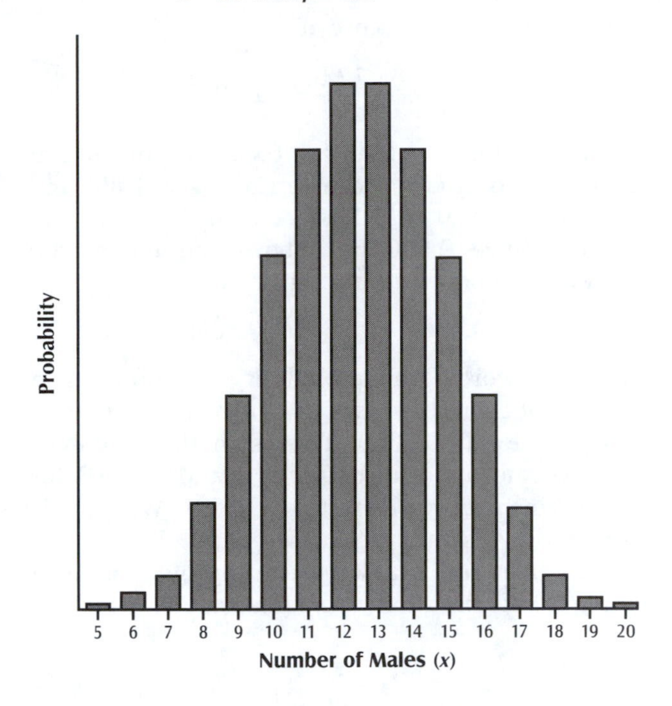


One approach is to use the binomial formula to calculate each of the nine probabilities ($x = 0, 1, 2, \ldots, 8$) and then sum each of these answers to arrive at our desired result, the probability of $x \leq 8$. This is tedious by hand. Using MINITAB, the answer comes more quickly: $P(x \leq 8) = .0322$. However, we can get the answer reasonably fast using the normal approximation to the binomial distribution.

The values of $k \times p$ and $k \times q$ are both equal to 12.5 (thus greater than 5), so we feel comfortable approximating this binomial distribution by using a normal distribution. The mean of a binomial variable is given by

$$\mu = kp \tag{6.3}$$

which, for the example, is

$$\mu = 25 \times 0.5 = 12.5$$

or 12.5 males out of 25 individuals.

The standard deviation is given by

$$\sigma = \sqrt{kpq} \tag{6.4}$$

which, for the example, is

$$\sigma = \sqrt{25 \times 0.5 \times 0.5} = 2.5$$

We may regard the possible outcomes of the various combinations (no males and 25 females, 1 male and 24 females, and so on) as a normally distributed variable with a mean of 12.5 and a standard deviation of 2.5. Reference to figure 6.9 might help to visualize this. The bars in figure 6.9 represent the binomial probability distribution for this example. Note that the general shape of this distribution is very similar to that of a normal distribution except that it is a bar graph. Equation 6.2 is now used to compute the $z$ score for a value of 8, given that the mean is 12.5 and the standard deviation is 2.5.

$$z = \frac{x - \mu}{\sigma}$$

The value 8 has a $z$ score of

$$z = \frac{8 - 12.5}{2.5} = -1.80$$

In table A.1 we find that 0.4641 of the normal curve lies between the mean and $z = 1.80$ (and therefore, that 0.4641 lies between the mean and $-1.80$). Since 0.500 lies above the mean, the region lying beyond a $z$ score of $-1.80$ is

$$0.5000 - 0.4641 = 0.0359$$

In other words, the probability of obtaining a sample of 25 individuals in which there are 8 or fewer males is 0.0359. This is in the neighborhood of the answer computed by MINITAB for exact binomial probabilities (.0322). When $n$ is larger, the answers are closer still.

To calculate a $z$ score for any outcome of a binomial distribution, use equation 6.5.

$$z = \frac{x - kp}{\sqrt{kpq}} \tag{6.5}$$

where $x$ is the number of cases in one category, whose probability is $p$; $kp$ is the mean of a binomial variable; and $\sqrt{kpq}$ is its standard deviation. This technique provides only an approximation of the binomial probability distribution, since the binomial distribution deals with discrete variables and the normal distribution deals with continuous variables. The discrepancy between the two is not large for large values of $k$ (30 or so), but it may become appreciable when $k$ is smaller.

## 6.5 Discrete Variables and the Normal Distribution

Many statistical tests assume that the variable in question has (approximately) a normal distribution. Since the normal distribution is a continuous probability distribution, we should ask whether it is appropriate to apply such a test to a discrete variable. As we saw in the preceding section, a discrete variable such as a binomial variable may be approximated by the normal distribution when $k$ (the number of cases or trials) is fairly large. This also holds true for a number of discrete variables. For example, the number of grapes in clusters varies from a few to almost 100. A variable such as this, although it is a discrete variable, might be approximately normally distributed in the same way that a binomial variable with a large $k$ is approximately normally distributed, and no serious harm is done when a statistical test that assumes approximate normality is used—provided, of course, that the other assumptions of the test are satisfied.

When the number of values that a discrete variable may assume is small, its distribution is very unlikely to be even approximately normally distributed, and the use of a statistical test that assumes approximate normality with such a variable is discouraged.

## 6.6 Parametric and Nonparametric Statistics

Many of the most commonly used statistical tests are based on some important assumptions about the nature of the variable in question. They assume that the variable is approximately normally distributed, that it is continuous (or if it is discrete, that it at least approximates a normal distribution), and that it is measured on an interval or ratio scale. When the data meet all of these basic assumptions, one may usually apply one of a group of tests known as **parametric tests**. Otherwise, one of another group of tests called **nonparametric tests** might be appropriate. Nonparametric methods are sometimes called distribution-free methods because they make few assumptions about the distribution of the variable in question. Nonparametric tests are also useful for data measured on an ordinal scale.

In later chapters, we explore the statistical analysis of some of the more common situations encountered in biological research. In each case a parametric test will be presented along with its nonparametric counterpart, when one is available. You are encouraged to pay particular attention to both the parametric and nonparametric alternatives and to the conditions under which each is to be used.

### Key Terms

nonparametric tests
normal distribution
parametric tests
standard normal distribution
$z$ scores

## Exercises

For exercises 6.1 through 6.4, regard the female mosquito fish in digital appendix 2 to be the entire population of interest. The mean length of this population is 34.29 mm and the standard deviation is 5.49 mm.

6.1 What is the probability that any individual selected at random from this population has a length of 50 mm or larger?

6.2 What is the probability that any individual selected at random has a length of 25 mm or shorter?

6.3 What is the probability that any individual selected at random has a total length of between 30 mm and 40 mm?

6.4 Determine what lengths include 95% (0.95) of the population.

For exercises 6.5 through 6.8, consider the bluegills in digital appendix 1 to be the entire population of interest. The mean standard length of this population is 152.09 mm and the standard deviation is 19.62 mm.

6.5 What is the probability that any individual sampled at random from this population will have a standard length of 170 mm or longer?

6.6 What is the probability that any individual sampled at random will have a standard length of 130 mm or shorter?

6.7 What is the probability that any individual sampled at random will have a standard length between 148 mm and 160 mm?

6.8 What lengths include 95% (0.95) of the population?

A population of red-bellied snakes is known to have a ratio of grey color morph to red color morph of 53:47. Use the normal approximation of the binomial distribution to solve exercises 6.9 through 6.14.

6.9 What is the probability of selecting from this population a random sample of 25 snakes containing 10 or fewer grey morph individuals?

6.10 What is the probability of selecting a random sample of 25 snakes containing 4 or fewer grey morph individuals?

6.11 What is the probability of selecting a random sample of 30 snakes containing 15 or fewer grey morph individuals?

6.12 What is the probability of selecting a random sample of 20 snakes containing 3 or fewer red morph individuals?

6.13 What is the probability of selecting a random sample of 50 snakes containing between 20 and 30 red morph individuals?

6.14 What is the probability of selecting a random sample of 40 snakes containing between 15 and 30 grey morph individuals?

6.15 Using the data from digital appendix 2, use statistical software (e.g., MINITAB) to construct a histogram and a probability plot of female mosquito-fish length. Use this information and any other procedures that your specific computer program has to decide if this variable is approximately normally distributed.

# 7

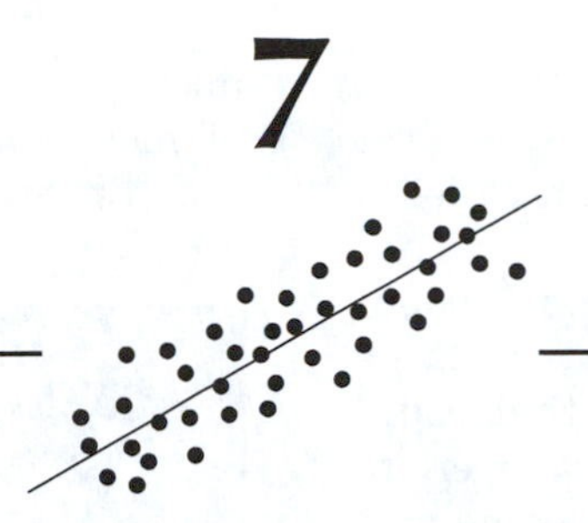

# Statistical Inference I

## Estimation and Sampling Distributions

## 7.1 An Introduction to Statistical Inference

It is very important for biologists to be able to make conclusions about populations based on samples taken from those populations. This activity is generally called **statistical inference**. There are two broad categories of statistical inference, illustrated by the following examples.

---

- **Example 7.1**

Estimating a population parameter

A random sample of 10 bluegill-sunfish standard lengths was selected. The mean of this sample was 159.40. We would like to know how well our sample mean estimates the population mean.

---

- **Example 7.2**

Testing a statistical hypothesis

Vitamin Y is an essential nutrient, but too much of it in the diet is harmful. Accordingly, the FDA sets standards for the vitamin Y content of vitamin pills: they must contain an average of 100 units of the vitamin per pill. The manufacturer of vitamin pills assigns a biochemist skilled in statistical inference to monitor the vitamin Y content of their product. She selects a random sample of 100 pills from a particular lot and finds that their mean vitamin Y content is 100.5 units, with a standard deviation of 2.19 units. She must decide if it is reasonable that a sample with a mean of 100.5 could have been drawn from a population with a mean of 100.

---

In example 7.1 we want to gain some idea about how well our sample mean, ($\bar{x}$), called a point estimator, actually estimates the population mean, $\mu$. This general category of statistical inference is called **estimation**. In example 7.2 we're testing the hypothesis that the average vitamin Y content of the manufacturer's pills is 100. Searching out answers to questions of this nature is generally called **hypothesis testing**. We examine estimation in this chapter. The theory behind hypothesis testing is introduced in chapter 8. Both topics are used repeatedly in subsequent chapters.

## 7.2 Estimating a Population Mean: The Central Limit Theorem

The sample mean in example 7.1 is 159.40. Based on this sample, we would like to conclude something about the population mean. Before we can do this, however, we must consider yet another type of probability distribution, called a sampling distribution.

When a random sample is taken from a population, that sample is only one of a very large number of samples of the same size that could have been taken (figure 7.1). In fact, if we were to take repeated random samples of the same size from a normally distributed population and compute the mean of each sample, we would find that not only are these means likely to be different

**Figure 7.1** Graphic representation of repeated samples from a population, such as the bluegill-sunfish example

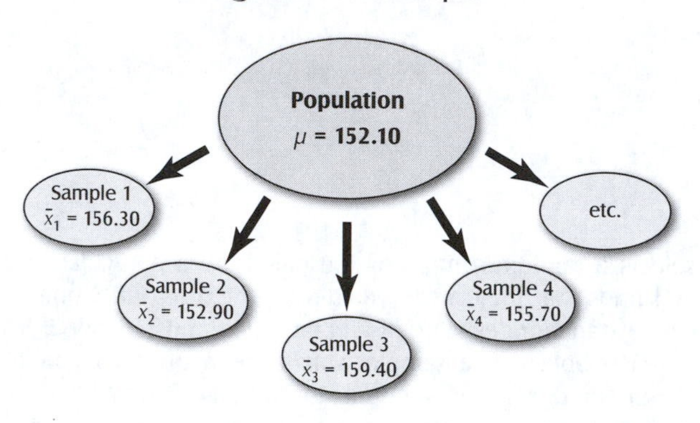


from each other but that they would have a range of values tending to cluster around some central value. In other words, these means would have a mean and standard deviation of their own and would exhibit all of the characteristics of a normally distributed random variable in exactly the same way that observations from a normally distributed population do. It is completely appropriate to think of a sample mean as an individual observation taken from a large population of possible sample means. The probability distribution that describes the behavior of a repeated sample statistic (mean, standard deviation, and so on) is called a **sampling distribution**. In the case of sample means taken from a normally distributed population with a known population standard deviation, the sampling distribution is the normal distribution. This seemingly simple concept is fundamental to much of statistical inference and may be stated formally as follows.

Figure 7.2 illustrates the central limit theorem. Here, the underlying distribution of city sizes is skewed, with most being small. Let's say we collect repeated samples of size $n$, compute the sample mean for each, and then plot the frequency distribution of the means. The distribution of means is less skewed and becomes approximately normal as the sample size gets larger.

**The Central Limit Theorem**

The means of samples from a normally distributed population have a normal distribution, regardless of sample size (n). The means of samples from a population with a nonnormal distribution have an approximately normal distribution as the sample size grows large.

**Figure 7.2** Demonstration of the central limit theorem. The original population is the population size (in thousands) of cities between 50,000 and 1 million people

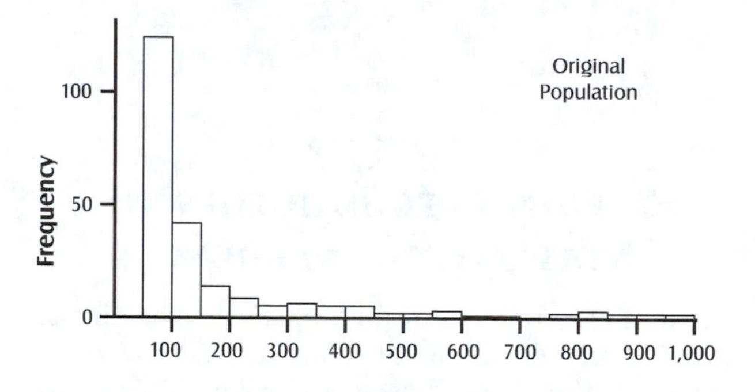


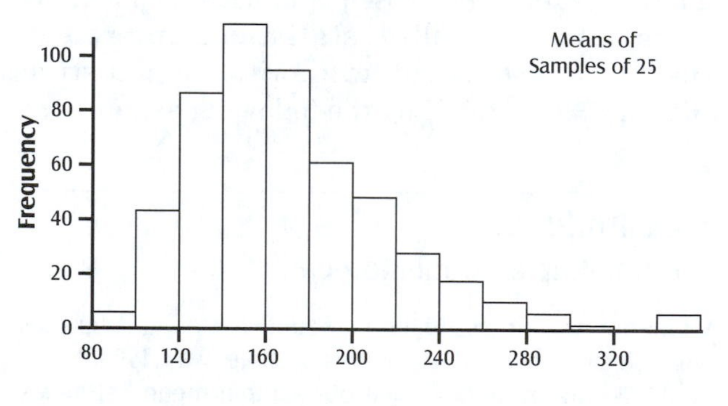


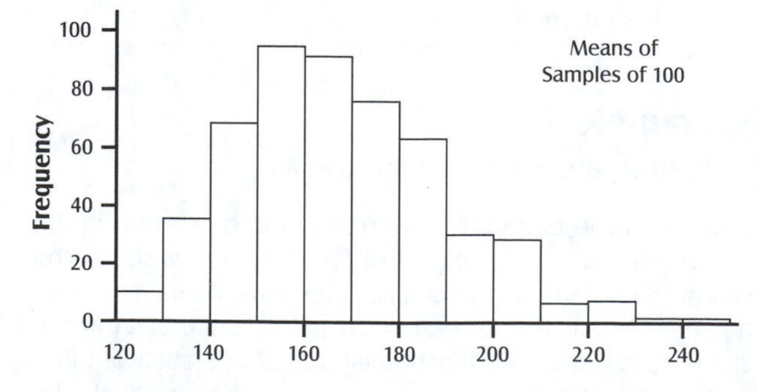


From Snedecor and Cochran 1980

Another important aspect of this sampling distribution is that as $n$ grows large, the distribution of means becomes narrower in shape, indicating that most sample means are close to the true mean $\mu$. In other words, our precision for estimating the mean (or any other parameter) improves as $n$ grows larger. This property is illustrated in figure 7.3. Sample means from large samples (bottom) have a greater chance of being close to the true mean than sample means from small samples (top).

**Figure 7.3** Sampling distribution of means from the same population

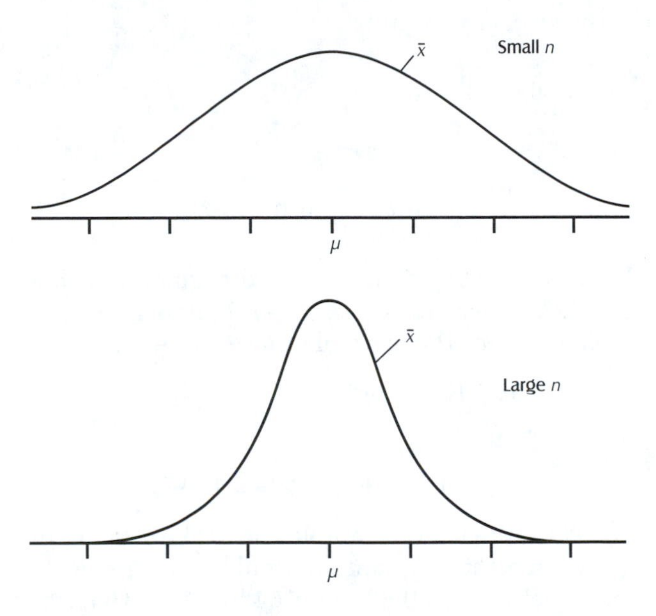


The useful thing about the central limit theorem is that we do not have to repeatedly take samples from a population to know how the means of such samples would behave if we did; they would have a normal distribution. Thus, we may consider a single sample mean from a normally distributed population (or from a non-normally distributed population, if our sample size is large) as one of many such possible means whose distribution would be normal.

## 7.3 Estimating a Population Mean: Standard Error of the Mean

We may assign a $z$ score to a sample mean if we know (or are willing to assume) the population mean ($\mu$) and standard deviation ($\sigma$) of the population from which our sample was taken. Recall (section 6.2) that $z$, a variant of the standard normal distribution, for a single observation is:

$$z = \frac{x - \mu}{\sigma} \tag{7.1}$$

### • Example 7.3

Assume that the bluegill sunfish whose lengths are given in digital appendix 1 are the entire population of interest. (Imagine that we had completely drained a pond.) The population mean length ($\mu$) for this population is 152.10 mm, and the population standard deviation ($\sigma$) is 19.64 mm. Suppose that we have taken a random sample of 10 individuals from this population (figure 7.1) and obtained a sample mean ($\bar{x}$) of 159.40 mm.

We may now ask what the probability is of obtaining a sample mean of 159.40 mm or larger from a population whose population mean is 152.10 mm and whose population standard deviation is 19.64 mm. We may answer this question in much the same way that we answered questions about the height of women students in example 6.1—that is, by assigning a $z$ score to this sample mean. This problem is illustrated graphically in figure 7.4.

**Figure 7.4** The standard normal distribution. Shaded area is the proportion above a $z$ score of 1.18

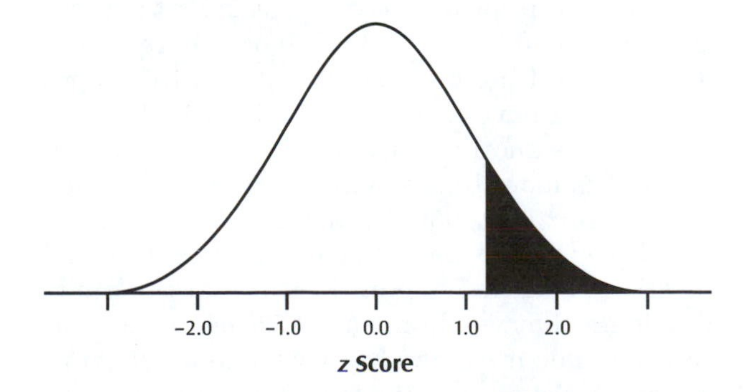

The $z$ score for a sample mean is given by

$$z = \frac{\bar{x} - \mu}{\sigma_{\bar{x}}} \tag{7.2}$$

The new term in this equation, $_{\bar{x}}$, is the standard deviation of our theoretical population of sample means; it is called the **standard error of the mean** and is calculated by

$$\sigma_{\bar{x}} = \frac{\sigma}{\sqrt{n}} \tag{7.3}$$

where $\sigma$ is the population standard deviation and $n$ is the sample size in question.

Notice how the standard error of the mean (equation 7.3) changes with sample size. As $n$ grows large, $_{\bar{x}}$ becomes small. This property is illustrated in figure 7.3. Sample means from large samples are more likely to be close to the true population mean than those of small samples. We have greater precision of our estimate.

For the example

$$\sigma_{\bar{x}} = \frac{19.64}{\sqrt{10}} = 6.21$$

The $z$ score for our sample is therefore

$$z = \frac{159.40 - 152.10}{6.21} = 1.18$$

We could, of course, calculate this more simply by combining equations 7.2 and 7.3 to give

$$z = \frac{\bar{x} - \mu}{\sigma / \sqrt{n}} \tag{7.4}$$

Consulting table A.1 we find that 0.3810 of the standard normal distribution lies between the mean and a $z$ score of 1.18. Subtracting this value from 0.5000 gives 0.1190 of the standard normal distribution that lies above a $z$ score of 1.18 (the shaded portion of figure 7.4). Thus, the probability of drawing a sample of 10 individuals with a mean as high as or higher than 159.40 mm is 0.1190. Another way of saying this is that approximately 12% (0.1190 × 100) of all of the samples of 10 that might possibly be drawn from this population would have a mean of 159.40 or higher.

There is another question we might ask, and one that is more to the point: Within what range would we expect 95% of the sample means of samples of 10 to lie? This is equivalent to the question asked in chapter 6 regarding the range of heights that would include 95% of the population of female freshman biology students. Refer to figure 7.5. We are looking for those values of sample means that include the unshaded portion of this curve, or the values

**Figure 7.5** The standard normal distribution. Shaded areas represent 0.05 of the distribution

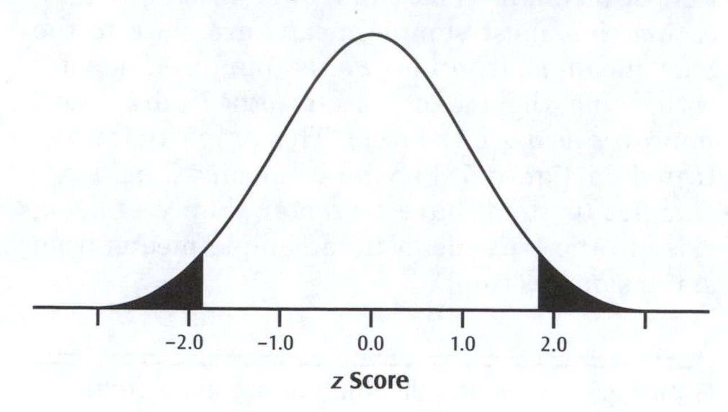


$$\mu + z_{(0.475)}\sigma_{\bar{x}} \tag{7.5}$$

and

$$\mu - z_{(0.475)}\sigma_{\bar{x}} \tag{7.6}$$

Recall that 0.95 of the standard normal distribution lies between $z$ scores of –1.96 and 1.96. (See table A.1.) For the example these values are:

$$152.10 + (1.96 \times 6.21) = 164.27$$

and

$$152.10 - (1.96 \times 6.21) = 139.93$$

Thus, 95% of the samples of 10 that we might draw from this population would have a sample mean between 139.93 and 164.27. Note that the sample mean we obtained for this example, 159.40, is within this range. Another way of thinking about this is as follows: there is a probability of 0.95 that any sample of 10 we might draw from this population would have a sample mean between 139.93 and 164.27; conversely, there is a probability of 0.05 that any sample mean would lie outside of this range—either smaller than 139.93 or larger than 164.27.

## 7.4 Confidence Interval of $\mu$ When $\sigma$ is Known

The information we have just considered is not very practical, since we would never want to estimate a population mean from a sample mean if we already know the value of the population mean. However, the concepts we have just examined provide the basis for a good deal of statistical inference, so it is important to grasp them. To see a practical application of this information, we change our assumptions about example 7.1 slightly. Suppose we do not know the population mean of this population but we do know the population standard error: 6.21. (Such information could have come from our prior experience in sampling bluegill populations.) Using the sample mean, we can calculate a range of values under which the population mean lies, with a specified level of confidence. This range is called a **confidence interval** of the mean. The procedure we follow in this case is applicable when the population standard deviation ($\sigma$) is known or can be assumed to have a particular value. A similar approach is used (section 7.5) when $\sigma$ is unknown.

**Assumptions of the Test**

1. The sample is a random sample from the population of interest.
2. Measurement is on an interval or ratio scale, and the variable is continuous.
3. The variable is approximately normally distributed.
4. The population standard deviation is known.

Using $\bar{x}$ as a point estimate of $\mu$ and armed with a knowledge of the population standard deviation, the upper limit (UL) and lower limit (LL) for a 95% confidence interval of the mean are given by

$$UL_{0.95} = \bar{x} + (1.96 \times \sigma_{\bar{x}}) \quad (7.7)$$

and

$$LL_{0.95} = \bar{x} - (1.96 \times \sigma_{\bar{x}}) \quad (7.8)$$

For the example:

$$UL_{0.95} = 159.40 + (1.96 \times 6.21) = 171.57$$

and

$$LL_{0.95} = 159.40 - (1.96 \times 6.21) = 147.23$$

Or we may combine these two expressions and write

$$95\% \text{ CI for } \mu = 159.40 \pm 12.17$$

or simply:

$$95\% \text{ CI for } \mu = 147.23, 171.57$$

Presenting this symbolically, the following statement is what we mean:

$$\Pr\{147.23 < \mu < 171.57\} = 0.95$$

Thus, we are 95% confident that the population mean for this bluegill population is included in the interval 147.23 to 171.57. Another way of saying this is that there is a 95% (0.95) probability that this range includes the population mean. To calculate a 99% rather than a 95% confidence interval for a mean, one simply uses 2.576 as the appropriate value for $z$, since this is the $z$ score that defines 99% (0.99) of the standard normal distribution—see table A.1.

**Caution**

When thinking about a confidence interval of a mean, or any other parameter, it is incorrect to conclude that there is a 95% (or 99% or whatever) probability that the parameter lies between the upper and lower limits. The parameter has a fixed value, usually unknown to us, and therefore there is no probability associated with it; it is whatever it is. The probability in such a case is that our confidence interval includes (or "encloses") the parameter in question.

To illustrate the meaning of a 95% confidence interval, imagine a population with known mean $\mu$ which we sample 20 times. For each sample, we compute a 95% confidence interval. We would predict that 19 of the 20 samples (95%) will correctly bracket $\mu$. This situation is illustrated in figure 7.6, with confidence intervals from individual

**Figure 7.6** Twenty 95% confidence intervals

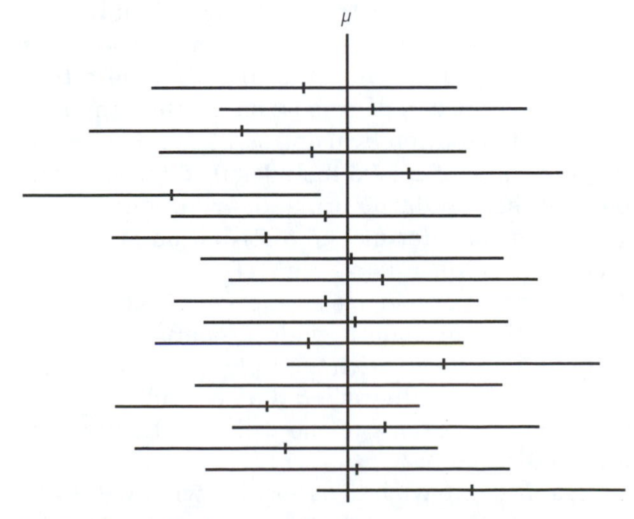


Adapted from J. Sumich, unpublished.

samples represented by horizontal lines. Notice that, although the sample means vary around $\mu$, all but one of the confidence intervals brackets $\mu$.

## 7.5 Confidence Interval of $\mu$ When $\sigma$ Is Unknown: The $t$ Distribution

In real life we usually do not know the population standard deviation; instead we must rely on the sample standard deviation as a point estimate of the parameter. Given a large sample from a normally distributed population, no serious error is committed by using the sample standard error as an estimate of the population standard error and using the standard normal distribution to compute confidence limits for a mean. However, it is not strictly correct to use the standard normal distribution. When the population standard deviation is unknown, the sampling distribution of means departs somewhat from normality, and this departure becomes more extreme as sample size becomes smaller. There is, however, a sampling distribution that better describes the distribution of means when the sample standard deviation must be used. This is called the $t$ distribution, or as it is sometimes known, Student's $t$ distribution. (By the way, "Student" was the pseudonym for William Gossett, a statistician employed by Guinness Brewery.)

Student's $t$ distribution resembles the normal distribution in most respects, except that it is defined by **degrees of freedom** $(n - 1)$ in addition to the mean and standard deviation. Recall that the proportion of the standard normal distribution that occurs between the mean and any $z$ score may be determined by reference to table A.1. We may also determine the value of $z$ that includes a specified proportion of the standard normal distribution by using table A.1. For example, a $z$ value of $\pm$ 1.96 includes 0.95 (or excludes 0.05) of the standard normal distribution.

When considering the distribution of sample means, it is useful to think of $t$ as having the same general properties as $z$, except that the values of $t$ that exclude a given proportion of the $t$ distribution vary with sample size. We may determine the value of $\pm t$ that excludes certain specified proportions of the $t$ distribution by reference to table A.2.

Suppose we wish to determine what value of $t$ excludes 0.05 of the $t$ distribution with 4 degrees of freedom. Figure 7.7 is a graphic representation of table A.2. The shaded portion of the curve is the proportion that lies outside of plus or minus any particular value of $t$. Keep this in mind while referring to table A.2. The column headings are the proportion indicated by the shaded areas in figure 7.7. Note that the proportions of the $t$ distribution given in table A.2 and figure 7.6 include both "tails" of the distribution. This is an important distinction between table A.1, the normal distribution, and table A.2. The left column (df) in table A.2 indicates degrees of freedom. The numbers in the body of the table are values of $t$ that correspond to a designated proportion of the distribution for the specified degrees of freedom. In the present case, the number corresponding to 0.05 and 4 degrees of freedom is 2.776, which is found in the table by following the 0.05 column down and the 4 df row across (refer to table A.2 now). Thus, $\pm$2.776 is the value of $t$ that excludes 0.05 of the $t$ distribution (or includes 0.95). This is shown graphically in figure 7.7.

**Figure 7.7** A $t$ distribution with 4 degrees of freedom

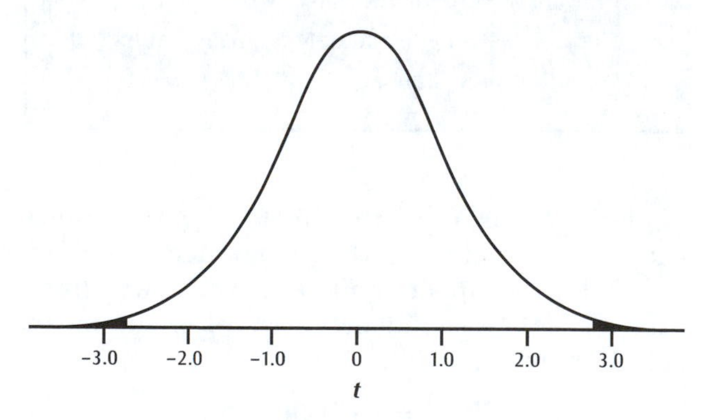


Suppose we wish to determine the value of $t$ that excludes only the upper 0.05 of the $t$ distribution with 4 degrees of freedom. Table A.2 gives $t$ values that exclude a certain proportion of the distribution equally divided between both tails, as shown in figure 7.7. Thus one-half of the indicated proportion is in the upper tail of the curve, and one-half is in the lower tail of the curve. Therefore, when we are interested in only one tail, we simply double the proportion indicated in table A.2. For this problem we consult table A.2 for 4 degrees of freedom and a probability (proportion) of 0.10 (rather than 0.05), where we find that 2.132 is the $t$ value that excludes the upper 0.05 of the $t$ distribution. This is shown graphically in figure 7.8.

**Figure 7.8** A *t* distribution with 4 degrees of freedom

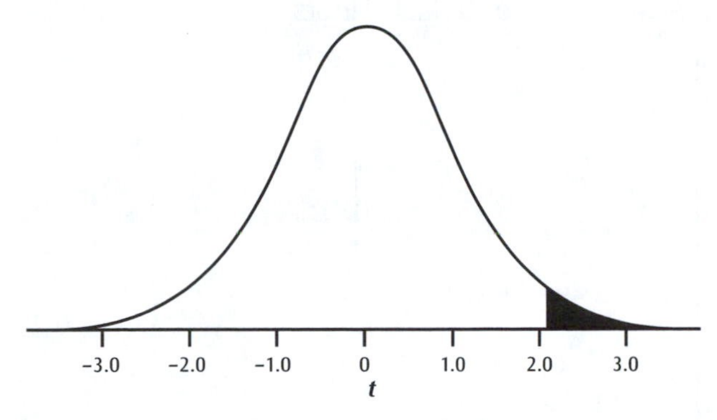


Note in table A.2 that $t$ for any particular proportion of the $t$ distribution decreases as the degrees of freedom increase. At infinite degrees of freedom (effectively, a large number > 120), the value of $t$ that excludes 0.05 of the distribution is 1.960—exactly the value of $z$ that excludes 0.05 of the standard normal distribution. This is because as degrees of freedom increase, the $t$ distribution tends to become a normal distribution. This property is useful, since the table of normal distribution areas (table A.1) allows us to determine probability more precisely than the $t$-table (table A.2), with its few columns of probabilities.

Returning now to our original purpose—determining the confidence interval for a mean when we must use the sample standard deviation—we use the $t$ distribution rather than the normal distribution as the appropriate sampling distribution. Otherwise, we use a similar approach as used earlier when $\sigma$ was known (section 7.4).

### • Example 7.4

A random sample of 20 male mosquito fish was collected, and total length (in millimeters) was determined. The sample mean length ($\bar{x}$) is 21.0 mm, and the sample standard deviation ($s$) is 1.76 mm. We wish to construct a 95% confidence interval for the population mean. Length in male mosquito fish is known to be approximately normally distributed.

---

This is a fairly typical situation. Both the mean and the standard deviation of the population must be estimated from their sample values, and the sample size is fairly small. The assumptions of the test are essentially the same as those for the construction of the confidence interval of a mean when the population standard deviation is known. They are repeated here.

**Assumptions of the Test**

1. The sample is a random sample from the population of interest.
2. Measurement is on an interval or ratio scale, and the variable is continuous (or, if discrete, the variable may assume a wide range of values).
3. The variable is approximately normally distributed.

The upper limit for a 95% confidence interval is given by

$$UL_{0.95} = \bar{x} + \left(t_{(0.05,\, n-1)}\right) \times s_{\bar{x}} \tag{7.9}$$

and the lower limit is given by

$$LL_{0.95} = \bar{x} - \left(t_{(0.05,n-1)}\right) \times s_{\bar{x}} \tag{7.10}$$

where $t_{(0.05,\, n-1)}$ is the tabular value of $t$ at $n - 1$ degrees of freedom, which delineates 0.95 of the $t$ distribution (found in table A.2 in the column headed "0.05" and the row corresponding to the appropriate degrees of freedom), and $s_{\bar{x}}$ is the sample **standard error of the mean,**

$$s_{\bar{x}} = \frac{s}{\sqrt{n}} \tag{7.11}$$

For example 7.4,

$$UL_{0.95} = 21.0 + \left(2.093 \times \frac{1.76}{\sqrt{20}}\right) = 21.825$$

and

$$LL_{0.95} = 21.0 - \left(2.093 \times \frac{1.76}{\sqrt{20}}\right) = 20.175$$

or

$$95\% \text{ CI for } \mu = 21.0 \pm 0.825$$

or simply

$$20.175 < \mu < 21.825$$

Thus, we conclude there is a 95% (0.95) probability that the range of 20.175 mm to 21.825 mm includes the population mean.

## 7.6 Reporting a Sample Mean and Its Variation

How should one report a sample mean in a scientific presentation? There are several more or less conventional ways of doing this, depending on the information one wishes to convey. The information in example 7.2 might be reported in any of the following ways.

1. The mean plus or minus the standard deviation $(\bar{x} \pm s)$, which for example 7.2 is 100.5 ± 2.19. While this tells the reader something about the variation of the measured variable in the population, it provides little information about how well the sample mean estimates the population mean, unless the reader wishes to calculate a confidence interval for him- or herself.
2. The mean plus or minus the standard error $(\bar{x} \pm s_{\bar{x}})$. This is perhaps the most common way of reporting a sample mean, and it does provide a rough idea of how well the sample mean estimates the population mean. However, it does not give this information as precisely as a confidence interval does.
3. The mean plus or minus the 95% (or 99%) confidence interval, or $\bar{x} \pm \text{CI}_{(0.95)}$. This method of reporting a sample mean has a great deal to recommend it. The information conveyed is directly accessible to the reader and is not misleading. Nevertheless, providing the standard error and the sample size allows readers to determine the CI themselves.

Sample means are often represented graphically, using one of the three methods just discussed. The most common form of such a graph is shown in figure 7.9. The points on the graph represent sample means, and the vertical lines, sometimes called "error bars," represent plus or minus one standard error of the mean.

### Key Terms

central limit theorem
confidence interval
estimation
hypothesis testing
sampling distribution
standard error of the mean
statistical inference

**Figure 7.9** Graphic representation of sample means and standard errors

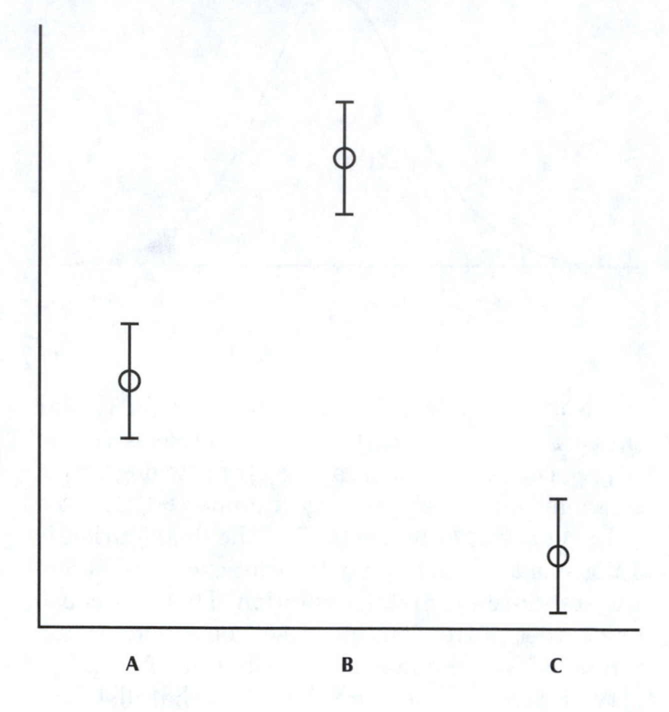


### Exercises

7.1 Compute the 95% confidence intervals for the samples of 10, 20, and 30 bluegill-sunfish standard lengths that you obtained in exercise 1.1. In this case assume the population standard deviation is known (41.95 mm) but the population mean is unknown.

7.2 Compute the 95% confidence intervals for the samples of 10, 20, and 30 male mosquito-fish lengths that you obtained from exercise 1.2. Assume the population standard deviation is known (2.64 mm) but the population mean is unknown.

7.3 Compute the 95% confidence intervals for the samples of 10, 20, and 30 resting pulse rates that you obtained from exercise 1.4. Assume the population standard deviation is known (13.51 bpm) but the population mean is unknown.

7.4 Compute the 95% confidence intervals for the samples of 10, 20, and 30 reaction times from exercise 1.5. Assume the population standard deviation is known (40.43 msec) but the population mean is unknown.

7.5 Combine your results from any of these exercises (7.1–7.4) with those of your classmates. Construct a graph like that shown in figure 7.6 for all your samples. Draw a vertical line on this graph that represents the population mean of the population you sampled. The population mean can be easily calculated using the digital appendix data set. Do all of the confidence intervals include the population mean? If not, explain why some might not.

7.6 Based on the number of students in your class, approximately how many of the confidence intervals from exercise 7.5 would you expect might not include the population mean?

7.7 Compute the 90%, 95%, and 99% confidence intervals for the population mean using the sample data from exercise 4.1. Neither the population mean nor the standard deviation are known in this case. Interpret your results: what is meant by any one of these confidence intervals? Which confidence interval is wider? Why?

7.8 Compute the 90%, 95%, and 99% confidence intervals for the population mean using the data from exercise 4.3. Neither the population mean nor the standard deviation are known in this case.

7.9 The weights (in milligrams) of spleens of 9 newly hatched turkeys are given below. Compute the 90%, 95%, and 99% confidence intervals for the population mean. (Data from F. McCorkel.)

| | | |
|---|---|---|
| 18.9 | 20.4 | 15.9 |
| 19.9 | 17.4 | 24.0 |
| 21.3 | 16.2 | 19.3 |

7.10 The 72-hour blastogenesis of chicken peripheral blood lymphocytes ($10^6$ cells/well) is given below. Compute the 90%, 95%, and 99% confidence intervals for the population mean. (Data from F. McCorkel.)

| | | |
|---|---|---|
| 236 | 314 | 471 |
| 305 | 414 | 616 |
| 304 | 414 | 301 |
| 225 | 369 | 402 |

7.11 The number of eggs produced in a single brood by 11 female green iguanas is given below. Compute the 90%, 95%, and 99% confidence intervals for the population mean. (Data from T. Miller.)

| | | |
|---|---|---|
| 33 | 50 | 46 |
| 33 | 53 | 57 |
| 44 | 31 | 60 |
| 40 | 50 | |

7.12 The activity of *o*-diphenol oxidase ($\mu lO_2$/mgP/min) was measured in 15 tomato plants. The results are given below. Compute the 90%, 95%, and 99% confidence intervals for the population mean.

| | | | | |
|---|---|---|---|---|
| 36 | 29 | 41 | 45 | 33 |
| 28 | 43 | 32 | 37 | 41 |
| 29 | 32 | 43 | 37 | 25 |

7.13 The snout vent length (in centimeters) in 25 newly born garter snakes selected at random from several litters are given below. Compute the 90%, 95%, and 99% confidence intervals for the population mean.

| | | | | |
|---|---|---|---|---|
| 6.5 | 4.3 | 4.6 | 6.0 | 4.7 |
| 6.2 | 5.8 | 5.4 | 5.2 | 4.8 |
| 4.9 | 5.0 | 4.7 | 3.4 | 3.9 |
| 5.1 | 5.4 | 4.8 | 3.8 | 6.1 |
| 4.8 | 3.9 | 3.8 | 6.1 | 5.2 |

7.14 The total weight (in grams) of 14 female iguanas are given below. Compute the 90%, 95%, and 99% confidence intervals for the population mean. (Data from T. Miller.)

| | | |
|---|---|---|
| 1450 | 2000 | 2000 |
| 1550 | 2435 | 2750 |
| 2200 | 1550 | 1800 |
| 1500 | 1050 | 2850 |
| 1650 | 2300 | |

# 8

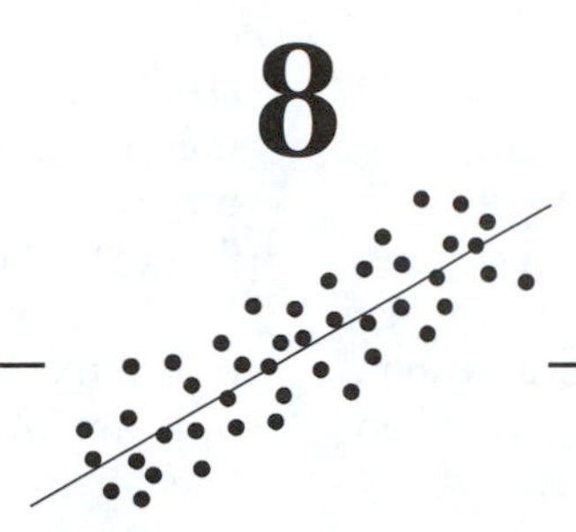

# Statistical Inference II

## Hypothesis Testing and the One-Sample *t*-Test

## 8.1 Statistical Hypothesis Testing and the Scientific Method

One useful definition of science goes something like this: "science is an organized body of knowledge about the physical universe, obtained by observation and experiment and used to make generalizations (theories) about the nature of the physical universe." There are some very important concepts embedded within this short definition, including why we need statistics!

Science begins with observations, which are simply events in the physical universe that we detect in some way. Measurement of a single individual in a population is an observation and its value is designated as $x$. Observations are important, but they tend to not be very useful in and of themselves. Science begins when we try to *explain* the observations. A tentative explanation of one or more observations is called a **hypothesis**; a good hypothesis has the following attributes.

***Attributes of a Scientific Hypothesis***

1. A hypothesis is consistent with the observations, which is to say that if it is correct, it will explain what has been observed.
2. If a hypothesis is false, it can be shown to be false. In other words, it can be tested. The test of a hypothesis is called an experiment.

Note that we do not say that if a hypothesis is true it can be shown to be true. Rather, we say that if it is false it can be shown to be false. Why such a negative attitude? Philosophers of science tell us that a false hypothesis may be proven to be false but a true hypothesis may not be proven to be true! Accordingly, a hypothesis is considered to be true when, by experiment (testing), it has not been proven false.

When efforts to prove a hypothesis false (experiments) fail to do so, our confidence in the correctness of the hypothesis increases. If the hypothesis has general or widespread application to events in the physical universe, we designate it as a **theory**. Like a hypothesis, a true theory cannot be proven to be true, but a false theory can be proven to be false. Lamarck's theory of the inheritance of acquired characteristics was a pretty good theory for its time—it simply failed to stand up to repeated testing (experiment), and so it was abandoned in favor of a better theory.

Science does not proceed by proving things—it proceeds by disproving things—and eventually, incorrect theories will be proven to be incorrect. This business of attempting to disprove hypotheses and theories is what we call the scientific method. The method operates by an *if. . . then* logic. *If* the hypothesis or theory is correct, *then* our prediction will be the outcome of the experiment. If the outcome of the experiment is something other than what the hypothesis or theory predicts, then we reject the hypothesis and look for a better explanation. This process is often called hypothesis testing, and it is the activity that distinguishes science from other forms of human knowledge-seeking.

Testing the conclusion(s) of a research endeavor by statistical means is a special application of the scientific method. It is often possible, particularly in hypothesis testing, to state a question or hypothesis in such a way that there are only two possible outcomes—"A is true" and "A is not true." Suppose that, unknown to us, the first of these statements, "A is true," is correct. We could not prove this statement directly if in fact it is correct because, as you will recall, a true hypothesis cannot be shown to be true. Suppose, though, that we tested the statement "A is false" and proved it to be incorrect (a false hypothesis can be shown to be false). If the statement "A is false" is disproved, and the statement "A is true" is the only other alternative, then we are left to conclude that the statement "A is true" is in fact correct!

Unfortunately, we must often rely on a sample of individuals from a larger population to obtain information about that population. There is always a probability that our sample is not really representative of the population. However, using hypothesis testing, we can specify a probability that our conclusion is incorrect. If this probability is low (by long-standing tradition, 0.05 or less), we feel confident that our conclusion is correct. There are special circumstances when investigators choose a higher or lower probability (say 0.10 or 0.01), but 0.05 is used most often and we will follow that proportion in this book.

In hypothesis testing, we usually attempt to reduce the question at hand to two choices, much like the "A is true" and "A is not true" dichotomy, and to specify exactly what we mean by these two choices. Only one of these statements is tested, and its parameter of interest is specified. The choice tested is called the **null hypothesis**, usually symbolized by $H_0$, and the other choice is called the **alternative hypothesis**, symbolized as $H_a$. Quite often, the alternative hypothesis is what we think is correct, what some call our working hypothesis.

An understanding of the nature of the null hypothesis is critical to understand the nature of hypothesis testing. Essentially, it is the hypothesis of no difference. ("Null," as an adjective, means, among other things, "amounting to nothing" or "having zero magnitude." It is derived from the Old French term *nul*, meaning "none.") Note carefully in the following examples that the null hypothesis always contains an equal sign, or at least an implied equal sign. This is because such a null hypothesis allows us to generate a probability distribution, which we demonstrate below. We do this by calculating a **test statistic**. A test statistic is a calculated value whose distribution is known when the null hypothesis is true.

## 8.2 Test of a Hypothesis Concerning a Single Population Mean: The One-Sample *t*-Test

Example 7.2 should help to clarify the concepts discussed above, and it introduces a few new ones. It might be useful to review the normal distribution (chapter 6) and the *t* distribution (section 7.5) in conjunction with this example. Refer back to example 7.2 now. The sample mean is 100.5 and the sample standard deviation 2.19.

In this case the population mean ($\mu$) is hypothesized to be 100 units, the specified FDA requirement. In this situation, the null hypothesis to be tested is that the mean vitamin Y content of the pills is 100 units, or

$$H_0\text{: } \mu = 100 \text{ units}$$

Note the equal sign in this expression. The alternative hypothesis (the one that is true if the null hypothesis is false) is

$$H_a\text{: } \mu \neq 100 \text{ units}$$

We assume, for the moment, that the population mean of our hypothetical population of vitamin pills is 100 units. We do not know if this is true; it is the population mean specified by the null hypothesis, and we wish to determine if it might be true. If we conclude that it is not true, then we conclude that the only other possibility—the alternative hypothesis—is true.

Recall from the discussion of sampling distributions of means (chapter 7) that there is a certain probability that a sample with a mean of 100.5 units could be drawn from a population with a population mean of 100 units by chance. Remember, we do not know the population mean. However *if* it is 100, *then* there is a certain probability associated with drawing a sample with a mean that differs by as much as 0.5 units from the population mean of 100. To test the null hypothesis, we calculate a $t$ statistic ($t_s$) by the following formula:

$$t_s = \frac{\bar{x} - \mu_0}{s/\sqrt{n}} \quad (8.1)$$

where $\mu_0$ represents our hypothesized value (in this example, 100), and we recognize $s/\sqrt{n}$ as the formula for the standard error of the mean (eq. 7.11). When the $t$ statistic is equal to or greater than the critical value of $t$ for a specified area of the $t$ distribution (usually 0.05), we reject the null hypothesis, since the probability of obtaining such a value of $t$ when the null hypothesis is true is quite small (0.05 or less). The critical value of $t$ is obtained from table A.2.

The assumptions of the $t$-test are the same as those for computing a confidence interval using the $t$ distribution (see section 7.5).

For the example,

$$t_s = \frac{100.5 - 100}{2.19/\sqrt{100}} = 2.283$$

the critical value of $t$ for $p = 0.05$ and 99 degrees of freedom (we use 100 degrees of freedom, since 99 is not tabulated) is 1.984 (see table A.2). Since the $t$ statistic (2.283) exceeds the critical value (1.984), we may reject $H_0$.

This problem is shown graphically in figure 8.1. The $x$-axis represents the distribution of $t$ that would be expected under the null hypothesis. The values +1.984 and −1.984 delimit 0.95 of the distribution. Thus, there is a probability of 0.95 that any sample we might take from this population would have a $t$ value of between −1.984 and +1.984, and a probability of 0.05 that any sample would have a value lower than −1.984 or greater than +1.984. Since our value of 2.283 is very unlikely if the null hypothesis is true, we reject the null hypothesis and conclude that the alternative hypothesis is true. Refer to figure 8.1 once again. The shaded areas—the proportion of the curve beyond ±1.984—are sometimes referred to as the region of rejection because when the $t$ statistic falls within this region, the null hypothesis is rejected. The null hypothesis is rejected because if it were true, we would expect, with a probability of 0.95, to obtain a calculated value of $t$ that falls within the unshaded portion of figure 8.1.

There are two explanations for obtaining a $t_s$ value as high as we did in this example. One is

**Figure 8.1** A $t$ distribution with 99 degrees of freedom for a two-tailed probability, $\alpha = 0.05$

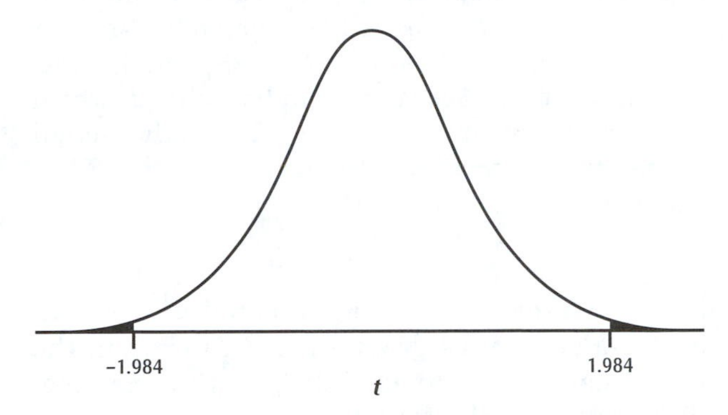


that the null hypothesis is true, and the sample mean we obtained differed from the population mean by as much as it did by chance alone. In other words, we obtained a very unlikely sample! The second explanation is that the null hypothesis is false. Since the probability of drawing a sample with a mean that differs this much from the population mean by chance alone is very small (less than 0.05), we conclude that the null hypothesis is false. Because both of these explanations are possible, we have to recognize that we can make a wrong decision in hypothesis testing. This topic is explored further in section 8.3.

As previously stated, $H_0$ is usually rejected when the probability that a sample value (such as a sample mean) would be obtained from a population with a population value (a mean, for example) specified by the null hypothesis is equal to or smaller than some predetermined level, such as 0.05. This value is referred to as **alpha** ($\alpha$). Note that alpha is not an exact probability associated with a particular value of the test statistic but rather a predetermined probability at or below which we conclude that the null hypothesis is false.

Ideally, the researcher determines what level of alpha he or she will regard as sufficient for rejecting $H_0$ before the experiment is conducted. In most cases, an alpha level of 0.05 is regarded as sufficient for rejecting a null hypothesis. Note that in the example above, the calculated value of $t_s$ (2.283) was considerably larger than the **critical value** of $t$ (the value that delimits the specified area of the distribution). Thus, we know that if the null hypothesis is true, the probability of obtaining a $t_s$ value with an absolute value as large as or larger than the value we obtained is actually less than 0.05. Table A.2 allows us to find a range for this probability, but not its exact value, since this table specifies values of $t$ for only certain probabilities. Consulting table A.2 at 100 degrees of freedom gives the critical value for a probability of 0.05 as 1.984, which is smaller than our calculated value of $t_s$ (2.283), and a critical value of 2.364 for a probability of 0.02, which is larger than our value of $t_s$. Thus, the actual probability associated with our value of 2.283 lies between 0.05 and 0.02, or

$$0.02 < p < 0.05$$

In this expression $p$ is the probability of obtaining a value of $t_s$ as great as or greater than the value that was obtained (2.283, for this example) if the null hypothesis is true. Such a probability is called the ***p* value**. For the example the $p$ value lies somewhere between 0.02 and 0.05, as determined by use of table A.2. Most statistical software gives exact $p$ values. For this example the $p$ value is 0.025, as determined by computer. When the $p$ value is equal to or lower than a predetermined value of alpha (usually 0.05), the null hypothesis is rejected. In this case, had we chosen to select an alpha level of 0.01, we would not reject the null hypothesis. Although alpha ($\alpha$) is customarily set at 0.05, there are occasions when alpha may be higher or lower. We will explore this concept further in section 8.4.

## 8.3 One-Tailed and Two-Tailed Hypothesis Tests

Note that in the above example we have included one-half of alpha, ($\alpha/2 = 0.025$) in each tail of the distribution (figure 8.1). (This area of a distribution used in hypothesis testing is sometimes called the "rejection region.") This is because the null hypothesis specified that $\mu$ = 100, and if rejected, it would tell us only that ≠ 100. Thus, we would reject $H_0$ if $t$ fell in either the upper or the lower tail of the distribution (i.e., $t \leq -1.984$, or $t \geq 1.984$). This is called a **two-tailed test** (also called a two-sided test). In two-tailed tests, it is the absolute value of $t$ that is important.

In some situations we have an interest in only one tail of a distribution, so these are referred to as **one-tailed tests**. This can be illustrated by changing our example slightly. Suppose the FDA requirement for vitamin pills specifies that pills must contain an average of *at least* 100 units of vitamin Y per pill. They may contain 100 units or more, but they may not contain less than 100 units. The null hypothesis for this case is

$$H_0: \mu \geq 100$$

and the alternative hypothesis is

$$H_a: \mu < 100$$

(Note that the null hypothesis contains an equal sign.)

In this situation we are only interested in the lower tail of the $t$ distribution. The shaded portion of figure 8.2 represents 0.05 of the $t$ distribution in this case, and the value of $t$ that delimits this area is –1.660. This critical value is found in table A.2 by looking in the column headed "0.1" at 100 degrees of freedom. (Table A.2 gives only two-tailed probabilities. To find a one-tailed probability, we must consult the column for twice the desired probability.) Thus, a $t_s$ value of –1.660 or lower would be justification for rejecting the null hypothesis. Since our calculated value for $t_s$ is much higher than this, we cannot reject $H_0$. We conclude that the vitamin pills contain 100 or more units of Y per pill. Note that this one-tailed test does not permit us to conclude that the pills contain exactly 100 units or more than 100 units; it only indicates that they do not contain less than 100 units. Note also that in a one-tailed test, the sign of $t$ is important.

**Figure 8.2** A $t$ distribution with 99 degrees of freedom for a lower one-tailed probability, $\alpha = 0.05$

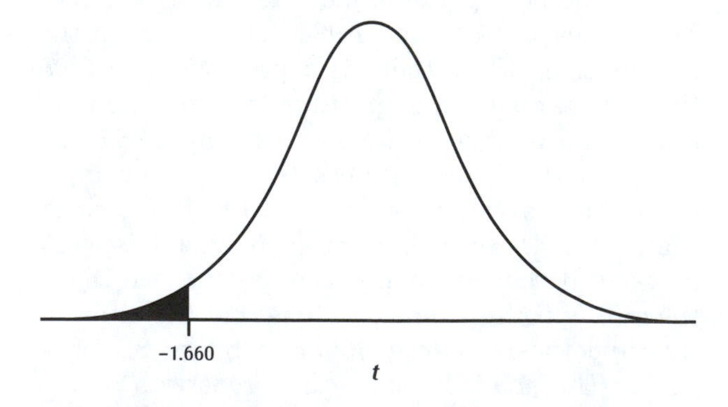


The same procedure is used when we wish to conduct a one-tailed test with alpha only in the upper tail of the distribution. In this case the null hypothesis would be that the pills contain no more than 100 units, or $H_0$: $\mu \leq 100$, and the alternative hypothesis is $H_a$: $\mu > 100$. In this case the critical value of $t$ is +1.660, and we would reject $H_0$ when our calculated value of $t$ is equal to or greater than this value. For the example we may reject $H_0$ and conclude that the pills contain more than 100 units. This situation is shown in figure 8.3.

**Figure 8.3** A $t$ distribution with 99 degrees of freedom for an upper one-tailed probability, $\alpha = 0.05$

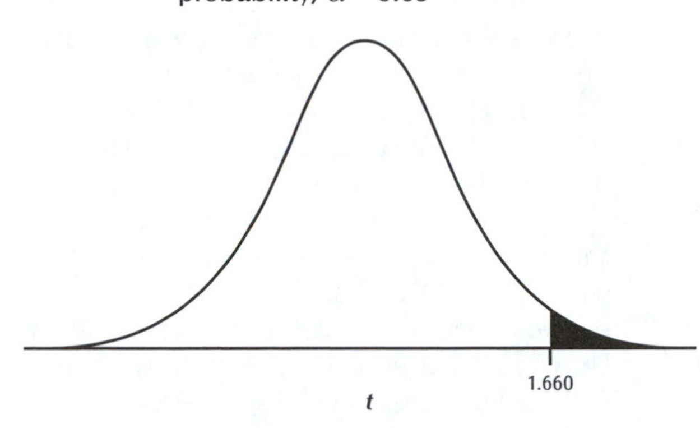


How does one know when a one-tailed test is appropriate and when a two-tailed test is appropriate? That depends on the null hypothesis, which in turn depends on the question we wish to answer. The three questions we asked concerning the vitamin-pill example are fairly typical. If we wish to know if the vitamin content of the pills is either more than or less than 100 units, a two-tailed test is indicated. On the other hand, if interest centers only in determining if the content is more than 100 units, we would choose a one-tailed test. Note, however, that we would not reject $H_0$ if the content of the pills were less than 100 units. Thus, it is very important to state the null hypothesis in such a way that, if it is rejected, we will arrive at a sensible answer to the research question we had in mind.

Let's consider one more example. One-tailed hypothesis tests are quite common when studying the effect of some experimental treatment. Imagine that we would like to know if a new drug is effective at lowering blood pressure. We could do an experiment by measuring the blood pressure on a group of subjects both before and after taking the medication. If the medication is effective, we would expect that the second blood-pressure readings would tend to be lower than the first readings. In particular, we would expect the mean of the differences (before – after, "b – a") would be greater than zero. We would thus create the following null and alternative hypotheses:

$$H_0: \mu_{b-a} \leq 0$$
$$H_a: \mu_{b-a} > 0$$

Clearly, evaluating this question requires a one-tailed hypothesis test. We would reject the null hypothesis only if the $t$ statistic were a large positive number, similar to the case shown in figure 8.3. We explore these sorts of questions about treatment effects in more detail in chapter 9.

**Caution**

Choice of a one-tailed or two-tailed hypothesis must come from the particular research question. It is never correct to first examine the data and then decide which hypothesis test is appropriate.

## 8.4 Statistical Decision Making and Its Potential Errors

Ideally, the researcher sets alpha before the statistical analysis is performed. In example 7.2 our biochemist set alpha at 0.05. Since the calculated value of $t_s$ was greater than the critical value for $t$, she rejected $H_0$. The actual probability associated with the calculated value of $t_s$ (the $p$ value) in this case was 0.025, as determined by computer, or between 0.02 and 0.05, as determined from table A.2. In either case $p \leq 0.05$, so $H_0$ was rejected. Suppose, however, that she had been willing to reject $H_0$ only if $p \leq 0.01$ (i.e., she had set alpha as 0.01). In this case she would have been unable to reject $H_0$! So what is the truth of the matter? Is $H_0$ true or false? The fact that it is possible to reach two contradictory conclusions from the same data, depending on where we set alpha, should convince you that there is always an element of subjectivity involved in statistical inference. (That's one reason why reporting the $p$ value is useful.) In fact, our biochemist cannot know with absolute certainty if the vitamin pills do or do not contain an average of 100 units of vitamin Y unless she analyzes every pill (i.e., the entire population), in which case there would be no pills left to sell!

You may have realized by now that there is a possibility of reaching an incorrect conclusion when that conclusion is based on a statistical test. This is indeed the case. In fact, two types of errors are possible in such a situation. In example 7.2, suppose the null hypothesis was actually true—that the average vitamin content of the pills was in fact 100 units—but the statistical test indicated otherwise. Remember that in this instance there is a probability of 0.05 or less of obtaining a $t_s$ value of 2.283 or higher when $H_0$ is true. This probability is so low that we would be "safest" in rejecting the null hypothesis. Such an error—rejecting the null hypothesis when it is in fact true—is called a **type I error**. The risk that such an error will occur is called **alpha**, and we set this ahead of time by choosing the critical value for the test. By reporting the *p* value, such as we did above, we know more information about the risk of making a type I error. (Suppose we obtained a *p* value of $p < 0.001$; we would be much more sure that the null hypothesis is incorrect!)

There is also another possible error—failing to reject a null hypothesis when in fact it is false. This is called a **type II error**, and its probability is called **beta** ($\beta$). Beta is often not determined in statistical tests. However, more scientists are seeing the value of making these calculations. We briefly visit this topic in chapter 9 (power of the test).

Table 8.1, sometimes called a "truth table," illustrates the relationship between the null hypothesis and type I and type II errors.

**Table 8.1** The possible consequences of a statistical decision

| | | Decision from the Test | |
|---|---|---|---|
| | | Fail to Reject $H_0$ | Reject $H_0$ |
| Reality | $H_0$ True | Correct | Type I Error |
| | $H_0$ False | Type II Error | Correct |

Important points about alpha ($\alpha$) and beta ($\beta$) that you should to keep in mind:

1. Alpha, as previously stated, is fixed by the investigator and/or the scientific community at a certain level, usually 0.05 (or sometimes 0.10 or 0.01). Therefore, its value is known. If the probability associated with our test is equal to or smaller than alpha, the null hypothesis is rejected.
2. Beta, on the other hand, is usually not calculated, and therefore its value is usually not known. For any given test, beta decreases as sample size increases. Thus, large samples result in a smaller beta risk than do smaller samples.
3. The beta risk varies among statistical tests. In some tests beta is higher, while in others it is lower. A test with a low beta risk has high **power**. Some tests are more powerful than others. Accordingly, when one has a choice of two or more tests for testing the same null hypothesis, the most powerful test should always be used.
4. Designating a lower level for alpha decreases the risk of a type I error (by definition), but at the same time it increases the risk of a type II error (beta). Suppose that in our vitamin-pill example, the null hypothesis of 100 units per pill is actually false. Using an alpha level of 0.05 leads to the correct conclusion. However, if alpha were set at 0.01, the null hypothesis would not be rejected, resulting in a type II error. One reason alpha is so frequently set at 0.05 is that this level represents a fairly reasonable compromise between type I error and type II error.

You have likely heard or read of results of a statistical test being referred to by **significance level**—that is, as being "significant" or perhaps even "highly significant." The expression "significant" usually means that $H_0$ may be rejected at an alpha level of 0.05 (i.e., the *p* value associated with the test statistic is 0.05 or less). "Highly significant" usually means that $H_0$ may be rejected when alpha is set at 0.01 (the *p* value is 0.01 or less).

This particular terminology is unfortunate, since most people who use the English language understand (correctly) that "significant" means approximately the same thing as "important," "meaningful," or "of some consequence." Thus, when some scientist informs you that his or her results are "highly significant," try not to be overly impressed! This simply means that he or she has very likely reached the proper conclusion about a question that might be important or that might be totally trivial. A better expression to use is "statistically discernable."

Fortunately, this particular terminology, and even the use of predetermined alpha levels, seems to be losing popularity in favor of simply expressing the *p* value associated with the test statistic. Thus, in example 7.2 our biochemist might report that $H_0$ was rejected at $p = 0.025$ (or $0.02 < p < 0.05$, if she used table A.2 rather than a computer) instead of reporting that the difference in vitamin content between her company's pills and the FDA requirement was "significant, but not highly significant" (which, of course, a normal person would interpret as "important, but not very important"). There are actually several ways that results of a statistical test might be reported, and you will encounter all of these in the biological literature. All mean more or less the same thing.

1. "The result is significant at the 95% level." (This number refers to the confidence level, $1 - \alpha$).
2. "The result is significant." By implication this means the same thing as the first statement.
3. "$p \leq 0.05$." This tells the reader that the $p$ value associated with the test statistic was equal to or less than 0.05, but there is no indication of how much less. The reader is free to decide for himself or herself if $H_0$ should be rejected.
4. "$0.02 < p < 0.05$." This tells the reader that the $p$ value of the test was less than 0.05 but not as small as 0.02. This method of reporting is even better than the immediately preceding example, since it gives the reader a much better indication of the actual $p$ value.
5. "$p = 0.025$." This statement tells the reader that the probability associated with the test statistic is 0.025 and lets the reader share with the investigator the decision about whether or not to reject $H_0$. Although most statistical tables do not give such exact probabilities, most computer programs for statistics do give exact probabilities. With the general availability of personal computers and statistical software, this method of reporting results is gaining popularity.

Of these various ways of reporting the same result, the last two are preferred. Not only does they avoid the use of words such as "significant" and "highly significant," which have unwanted, misleading, and unavoidable connotations, but it also lets the consumer of scientific information decide what level of alpha is acceptable to him or her.

## 8.5 Steps in Testing a Hypothesis

Most hypothesis testing more or less follows a particular sequence of steps. In solving the exercises at the end of this and subsequent chapters, you should follow this sequence.

1. State, very clearly, the question you are attempting to answer.
2. Identify the characteristics of the sample and the variable in question. Is the scale of measurement nominal, ordinal, ratio, or interval? Is the variable known to be (or may it be assumed to be) approximately normally distributed? If not, what is the distribution?
3. What sampling distribution describes a sample of this kind, and what is the appropriate statistical test?
4. Based on your answers to numbers 1, 2, and 3 above, state the null hypothesis and the alternative hypothesis ($H_a$). Is a one-tailed or a two-tailed test appropriate?
5. Determine the level of alpha at or below which you will reject the null hypothesis, and locate the critical value from the appropriate table. (This is sometimes referred to as the decision rule.)
6. Make the appropriate calculation of the test statistic.
7. Using the appropriate table, determine the $p$ value.
8. Make your decision about $H_0$: if the probability of obtaining this calculated value is equal to or smaller than the pre-selected value of alpha, reject the null hypothesis and accept the alternative hypothesis.
9. Interpret our decision in light of the original question (#1 above).

Along with doing a hypothesis test, one should always provide some additional supporting evidence, such as a graph or table of descriptive statistics (chapters 3 and 4). This allows you (and a reader) to visualize whether or not your answer makes sense. If your graph shows that a mean is clearly different from a hypothesized value and your test fails to reject, then you may have made an error somewhere.

### Key Terms

alpha
alternative hypothesis ($H_a$)
beta
critical value
hypothesis testing
null hypothesis ($H_0$)
one-tailed test
$p$ value
power
significance level
statistical inference
test statistic
two-tailed test
type I error
type II error

## Exercises

8.1 The FDA has established that the concentration of a certain pesticide in apples may not exceed 10 ppb. A random sample of 100 apples from a major orchard had an average pesticide content of 10.03 ppb with a standard deviation of 0.12 ppb. Are the apples within the FDA requirement?

8.2 A certain enzyme in the liver of fish is considered an indicator of trace amounts of a dangerous pollutant that is difficult to detect by chemical methods. Enzyme activities of less than 50 units per gram of liver (fresh weight) are taken to indicate the presence of the pollutant. A random sample was taken from a local stream and the enzyme concentrations were as follows. Are enzyme activities depressed in this population?

| | | | |
|---|---|---|---|
| 48 | 43 | 51 | 42 |
| 50 | 42 | 44 | 45 |
| 56 | 49 | 44 | 47 |
| 50 | 49 | 38 | 46 |
| 38 | 52 | 32 | 56 |

8.3 A turkey geneticist wished to select for breeding purposes hens whose eggs have an average weight of 100 grams. The mean weight of a sample of 20 eggs from one hen was 97 grams, with a standard deviation of 2.5 grams. Should this hen be kept in the breeding program?

8.4 Healthy populations of buffobirds have newborn chicks that weigh, on average, 8.0 kg. We expect that chicks from populations impacted by disease to weigh less. Suppose we visit a new population and sample 20 randomly selected chicks. Use a $t$-test (at $\alpha = 0.05$) to test whether this population has chicks that are normal or below normal in weight.

8.1 8.0 8.0 7.7 8.1 7.9 7.6 7.6 7.9 8.1
7.8 8.2 8.1 8.3 8.0 8.0 7.8 7.9 7.6 8.1

8.5 Suppose that the mean water hardness of lakes in Kansas is 425 mg/L and these values tend to follow a normal distribution. A limnologist would like to know whether stock ponds tend to have lower hardness. He collected water from 25 randomly selected ponds, which yielded the following results. Test the appropriate null hypothesis (at $\alpha = 0.05$).

346 496 352 378 315 420 485 446 479 422
494 289 436 516 615 491 360 385 500 558
381 303 434 562 496

8.6 Suppose that the manufacturer of UltraChic cigarettes claims that its average nicotine content does not exceed 3.5 mg per cigarette. Based on prior studies, we know that nicotine content tends to be normally distributed. We collect a random sample of 10 cigarettes, make measurements in an independent laboratory, and determine the sample mean to be 4.2 mg, with a standard deviation of 1.4 mg nicotine per cigarette. Are these data consistent with the manufacturer's claim?

# 9

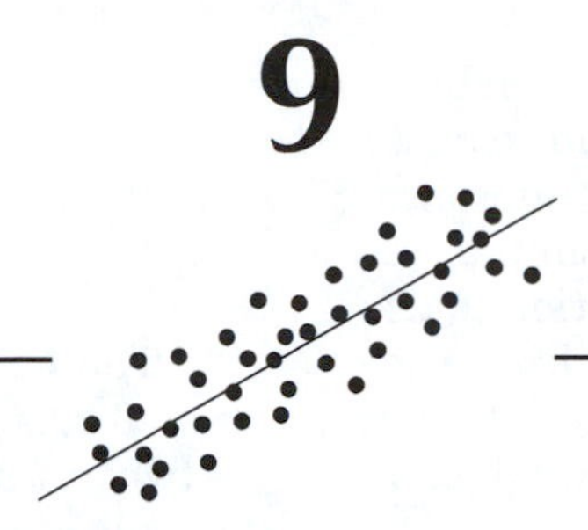

# Inferences Concerning Two Populations and Paired Comparisons

One of the more commonly used groups of statistical tests is that designed to test whether two or more populations differ from each other in some way. How does resting heart rate differ between males and females? Between smokers and nonsmokers? In both these cases, we would survey different populations. How does treatment with copper sulfate affect the density of mucus cells on the gills of bluegill sunfish? Does a new type of inhaler provide more relief to people suffering from asthma than the old type of inhaler? In these two cases we could take a large group of individuals (sunfish or people with asthma) and randomly assign samples to two different treatments. We are most often interested in differences in the means of these populations, but we can also test for differences in variability or distribution. This chapter considers methods for dealing with questions about whether two population means differ. Chapter 10 considers procedures for testing hypotheses about more than two population means.

In the first part of this chapter, we deal with methods of determining if differences in the means of two **independent samples** represent real differences in the two population means. In this context *independent* means that the two samples are from different populations. In the second part of this chapter, we deal with **related samples**, or as they are often called, matched-pairs samples. In related-samples experiments, each individual is measured twice or carefully matched pairs of individuals are measured.

When its assumptions are met, the most powerful test for comparing two independent populations is the two-sample *t*-test. The *t*-test is an example of a parametric test. A very useful nonparametric test that can usually be used when the data do not meet the necessary assumptions is the Mann-Whitney test. Both tests are discussed in the following sections.

## 9.1 The *t*-Test for Two Independent Samples

Recall from the previous discussion of sampling distributions (chapter 7) that the means of samples from a population of unknown variance have a $t$ distribution if the measured variable is at least approximately normally distributed. We used this concept in chapter 8 to make inferences about a single population mean. By a similar line of reasoning, we may use the $t$ distribution to test whether two population means differ from each other. The null hypothesis of such a test is that the means of the two populations are equal, or

$$H_0: \mu_a = \mu_b$$

If repeated sample means from a population have a $t$ distribution, then it follows that the difference between repeated *pairs* of sample means taken from a population or from two populations with the same population mean have a $t$ distribution.

$$t = \frac{(\bar{x}_a - \bar{x}_b) - (\mu_a - \mu_b)}{s_p} \quad (9.1)$$

where $(\bar{x}_a - \bar{x}_b)$ is the difference between the two sample means and $(\mu_a - \mu_b)$ is the hypothesized difference between the two population means. When the null hypothesis is that the two population means are equal (the most common case), $(\mu_a - \mu_b)$ is zero. The term $s_p$ is a standard error for the difference between means and is based on the pooled estimate of variance of the two samples and it is calculated as shown in the denominator of equation 9.2.

**Assumptions of the Test**

1. Samples are collected randomly from the two populations of interest.
2. The measured variable is approximately normally distributed and is continuous. If the measured variable is discrete, then it must assume a large range of possible values (see section 6.4).
3. Measurement is on an interval or ratio scale.

The $t$-test for independent samples is both powerful (meaning that the probability of a type II error is not high) and robust (meaning that it is still valid when the characteristics of the data depart somewhat from the assumptions). This is especially true when sample sizes are large. However, the power of the test becomes poor when the failure of the assumptions is severe. When assumptions 2 or 3 are not met, the Mann-Whitney test (section 9.3) might be appropriate. To conduct the two sample $t$-test, the test statistic $t$ is computed by equation 9.2.

$$t = \frac{(\bar{x}_a - \bar{x}_b) - (\mu_a - \mu_b)}{\sqrt{\dfrac{s_a^2}{n_a} + \dfrac{s_b^2}{n_b}}} \quad (9.2)$$

The expression in the denominator of equation 9.2 is the standard error of the difference between the two means. The degrees of freedom for this case are $n_a - 1$ or $n_b - 1$, whichever is the smaller of the two.

## • Example 9.1

A $t$-Test

The ratio of length to width of root hair cells in two species of plants (A and B) of the same genus were measured using random samples. Assume this variable is approximately normally distributed. We wish to know if the two population means differ. The results were as follows.

| | Species A | Species B |
|---|---|---|
| n | 12 | 18 |
| $\bar{x}$ | 1.28 | 4.43 |
| $s^2$ | 0.112 | 7.072 |

Data from T. Ruhlman

The null hypothesis for this example is

$$H_0: \mu_A = \mu_B$$

and the alternative hypothesis is

$$H_1: \mu_A \neq \mu_B$$

Because of the way the question is stated (do the two population means differ?) this is a two-tailed test. Substituting the example values in equation 9.2 gives

$$t = \frac{1.28 - 4.43}{\sqrt{\dfrac{0.112}{12} + \dfrac{7.072}{18}}} = -4.967$$

Since this is a two-tailed test, we are interested only in the absolute value of our calculated $t$, which gives us 4.967. The critical value of $t$ for 11 degrees of freedom and $\alpha = 0.05$ is 2.201 (see table A.2). The $p$ value is $0.0001 < p < 0.001$. We therefore reject $H_0$ that the two population means are equal and conclude that they are not equal.

Statistical software may sometimes present more than one version of the $t$ statistic. Another version of the $t$ statistic can be calculated when we assume the two population variances are equal. Since the procedures for testing equality of variance for small samples are not very powerful, many researchers prefer to use the method described above. The formulas we present for computing the pooled estimate of the variance and degrees of freedom provide the most conservative approach for conducting the *t*-test. In other words, this approach makes it more difficult to reject the null hypothesis.

## 9.2 Confidence Interval for the Difference between Two Population Means

Earlier (section 7.5) we calculated and interpreted a confidence interval for a single population mean. The same approach is also useful for determining the interval that likely brackets the difference between two means. The 95% confidence interval for $\mu_a - \mu_b$ is:

$$(\bar{x}_a - \bar{x}_b) \pm (t_{.05,k}) \times \sqrt{\frac{s_a^2}{n_a} + \frac{s_b^2}{n_b}} \qquad (9.3)$$

where $k$ is the degrees of freedom, the smaller of $n_a - 1$ or $n_b - 1$. Substituting the root hair statistics in equation 9.3 (and reversing the order of the groups to give a positive difference) gives:

95% CI for $\mu_a - \mu_b$

$$= (4.43 - 1.28) \pm 2.201 \times \sqrt{\frac{7.072}{18} + \frac{0.112}{12}}$$

$$= 3.15 \pm 2.201 \times \sqrt{0.4022}$$

$$= 3.15 \pm 1.40 = 1.75,\ 4.55$$

We are 95% sure that the mean difference between groups B and A in their root hair cell length-to-width ratios is at least 1.75 and as great as 4.55. We now know more than that the two groups are different in their root hair characteristics. We also know the magnitude of this difference, as well as our uncertainty about the difference.

Computer statistical software can easily be used to do two-sample *t*-tests and compute their confidence intervals. An example from MINITAB follows, using the mosquito-fish data (table 9.1).

Interpret this example on your own. Which group appears to be larger? (Inspect the sample means.) By how much? (Inspect the confidence interval and say that the difference is "at least ___ and as much as ___.") What is meant by a 95% confidence interval?

**Table 9.1** Comparison of male and female mosquito fish in their mean length (data from digital appendix). 1 = male, 2 = female

**Two Sample T-Test and Confidence Interval**

```
Two sample T for MosFishLength
Gender    N       Mean      StDev     SE Mean
1         854     23.60     2.64      0.090
2         797     34.29     5.50      0.19

95% CI for mu (1) - mu (2): ( -11.115, -
10.27)
T-Test mu (1) = mu (2) (vs not = ): T = -
49.83 P = 0.0000 DF = 1126
```

## 9.3 A Nonparametric Test for Two Independent Samples: The Mann-Whitney Test

There are a number of situations in which the data collected in a survey or experiment do not meet the assumptions of the *t*-test for independent samples. For instance, measurement might be on an ordinal scale or the distribution of the measurement variable might not be normal. In such cases one is well advised to use the Mann-Whitney test, which is the nonparametric counterpart of the *t*-test for two independent samples. When the assumptions of the *t*-test are not met, the Mann-Whitney test is more powerful. On the other hand, when the assumptions of the *t*-test are met, the *t*-test is more powerful (hence why we don't just always use the Mann-Whitney test!).

The Mann-Whitney test examines whether two samples could have been drawn from identical populations. Specifically, it tests whether two populations of the same but unspecified distribution differ with respect to central tendency. We'll typically use the median ($\theta$) for our measure of interest, since it works well with skewed distributions (section 4.2). The null hypothesis of the Mann-Whitney test is that the samples were drawn from populations with an identical median. In other words:

$$H_0\text{: } \theta_1 = \theta_2$$

The assumptions of the test are that the two population distributions are of the same shape (but not necessarily normal) and that random samples have been drawn from the two populations. The sample sizes need not be equal.

To conduct the Mann-Whitney test, the data of the two samples are ranked together, while at the same time identity of the sample to which each datum belongs is preserved. The lowest value in either sample receives a rank of 1, the next lowest the rank of 2, and so on. Tied scores receive the average rank that each would have, had they not been tied. It is not as complicated as it sounds. Example 9.2 will help clarify this. To aid in ranking, we first sorted the data in each group.

## • Example 9.2
### The Mann-Whitney test

Male gully cats are territorial; they hold territories up to several hectares in size. The territory size of random samples of gully cats from 2 locations was measured (in hectares) with the results shown as follows. We wish to know if there is a difference in territory size between these 2 populations. A glance at the raw numbers suggests that the average (median) territory size at location B is greater than that at location A. But we must run a formal hypothesis test to support our intuition.

| **Location A** | | **Location B** | |
|---|---|---|---|
| Territory size | Rank | Territory size | Rank |
| 7 | 1.5 | 8 | 3 |
| 7 | 1.5 | 10 | 4.5 |
| 10 | 4.5 | 18 | 8 |
| 14 | 6 | 21 | 11 |
| 17 | 7 | 29 | 14 |
| 20.6 | 9 | 32 | 15 |
| 21 | 11 | 35 | 16 |
| 21 | 11 | 36 | 17 |
| 24 | 13 | 37 | 18 |
| | | 45 | 19 |
| $n_a = 9$ | $\Sigma R_a = 64.5$ | $n_b = 10$ | $\Sigma R_b = 125.5$ |

Why might we choose the Mann-Whitney test for these data rather than a *t*-test for independent samples? We use this nonparametric test because we suspect that territory size in these imaginary animals is not normally distributed. Furthermore, because the sample sizes are small, the central limit theorem does not help us out.

First, all of the observations in both samples are ranked, with tied observations receiving the average rank that they would have if they were not tied. Note in the "location A" sample that there are two observations of 7, and these are the lowest of all of the observations. If they were not tied, they would receive the ranks 1 and 2. However, since they are tied, they receive the average rank that they otherwise would have, or (1 + 2)/2 = 1.5. Ranking proceeds in this manner until all observations have been ranked.

We now compute 2 values of the test statistic, *U*, by the following equations:

$$U_a = n_a n_b + \frac{n_a(n_a+1)}{2} - \Sigma R_a \tag{9.4}$$

where $n_a$ and $n_b$ are the sample sizes of samples A and B, and $\Sigma R_a$ is the sum of the ranks of sample A. $U_b$ is then calculated by

$$U_b = n_a n_b - U_a \tag{9.5}$$

For the example

$$U_a = (9 \times 10) + \frac{9(9+1)}{2} - 64.5 = 70.5$$

and

$$U_b = (9 \times 10) - 70.5 = 19.5$$

Table A.4 gives the critical values of *U* when both sample sizes are smaller than 20. The table is divided into two parts. For two-sided tests (at $\alpha = 0.05$), use the top half; for one-sided tests (at $\alpha = 0.05$), use the bottom half. The null hypothesis is rejected if either $U_a$ or $U_b$ is equal to or larger than the critical value. In this example, $U_a$ (70.5) is the larger of the calculated values. Checking table A.4 for a two-sided test at $\alpha = 0.05$ and sample sizes of 9 and 10, the critical value of *U* found in table A.4 is 70. Since our larger calculated *U* is larger than this, we reject $H_0$ for the gully cat example and conclude that territory sizes in the two locations are not equal.

Because of space limitations, table A.4 has critical values for a limited range of alpha. We thus cannot determine *p* values very precisely. However, the fact that the test statistic (70.5) was just slightly larger than the critical value (70) suggests that the *p* value is just less than 0.05. Statistical packages, such as MINITAB, report an exact *p* value. Table 9.2 shows the results from MINITAB for this same data set. The test statistic that is reported (*W*) is simply the sum of the ranks for the first group. Notice that the *p* value (0.0407) is right where we expect, less than 0.05, but not by much.

To conduct a *one-sided test*, we do two things differently from the two-sided test. First, at $\alpha = 0.05$, we use the bottom half of table A.4. Second, we use either $U_a$ or $U_b$, depending on the following criteria, which follow from the structure of the hypotheses:

**Table 9.2** Mann-Whitney test with MINITAB. Data from example 9.3

**Mann-Whitney Test and CI: LocationA, LocationB**

```
              N    Median
LocationA     9    17.00
LocationB    10    30.50
Point estimate for ETA1-ETA2 is -12.50
95.5 Percent CI for ETA1-ETA2 is
(-22.01,0.01)
W = 64.5
Test of ETA1 = ETA2 vs ETA1 not = ETA2 is
significant at 0.0412
The test is significant at 0.0407
(adjusted for ties)
```

If $H_0$: $\theta_a \geq \theta_b$ (vs. $H_a$: $\theta_a < \theta_b$), then use $U_a$ (from eq. 9.4) for the test statistic.

If $H_0$: $\theta_a \leq \theta_b$ (vs. $H_a$: $\theta_a > \theta_b$), then use $U_b$ (from eq. 9.5) for the test statistic.

Whenever conducting a one-sided test, it is always a good idea to inspect the data to make sure the result makes sense. If you reject the null hypothesis and conclude that one group has a larger median than a second group, then that first group should tend to have larger values than the second group. Plot your data and use descriptive statistics to visualize this.

Note that table A.4 applies only when sample sizes are 20 or less. If each sample has more than 20 observations, the probability distribution of $U$ is approximately normal. This allows us to calculate $z$ and use the table for the normal distribution (table A.1). To calculate $z$, use equation 9.6 below:

$$z = \frac{\left[U - \left(\frac{n_a n_b}{2}\right)\right]}{\sqrt{\frac{n_a n_b (n_a + n_b + 1)}{12}}} \quad (9.6)$$

$U$ is either $U_a$ or $U_b$, calculated according to equations 9.4 and 9.5. The denominator in equation 9.6 is the standard error of $U$. $H_0$ is rejected at = 0.05 if $|z| \geq 1.96$ for a two-sided test.

## 9.4 Tests for Two Related Samples

It is sometimes advantageous to use paired samples rather than independent samples, particularly in experimental work where the choice of subjects is under the control of the investigator. In certain instances an individual may even be used as its own control, such as in a before-after study. In other instances, individuals that are as nearly alike as possible may be selected to receive two or more different experimental treatments. For instance, pairs of potted plants may be distributed at different locations in a greenhouse. Using paired samples, it may be assumed that any difference between the groups is associated with the treatment given to them. This avoids a common experimental problem—ensuring that the control group is as nearly like the experimental group as possible except for the treatment given to the experimental group.

Suppose we wish to test the effect of some drug on pulse rate in humans. We might randomly assign individuals to two groups and give one group (the control group) a placebo and the other group (the treatment group) the experimental drug. Suppose also that the individuals in our study differed with respect to age, sex, smoking, caffeine consumption, and exercise habits. This would introduce a great deal of variation among the individuals unrelated to the treatment. We assume that, since individuals were randomly assigned to the two groups, the variance of the two groups would be equal. This variation from individual to individual that is not related to the treatment is called **error variance**. (This concept is discussed in more detail in chapter 10.)

Recall that the test statistic, $t$, is computed by

$$t = \frac{\bar{x}_A - \bar{x}_B}{s_p}$$

if we want to test the null hypothesis that there is no difference in population means between two independent samples. If the difference between the two population means is fairly small and the error variance is large, we would run a fairly high risk of failing to reject the null hypothesis when in fact it is false. In other words this situation would be likely to produce a type II error. This risk could be reduced by using a very large sample, but there is a much more efficient way.

Suppose we paired or matched the subjects in the study, pairing individuals that are as nearly alike as possible, and then we randomly assigned one member of each pair to one group (treatment or control) and the other member of the pair to the other group. Most of the variation between two individuals of a pair would presumably be caused by the difference in treatment given them since, in most other respects, they are alike. Thus, in a paired design, our interest focuses on the *differences* between pairs, and the tests conducted are essentially one-sample tests on the differences. This will become clear as we proceed.

This kind of test is widely used and goes by several names: paired comparison, matched pairs, and before-and-after study. A related design is called a randomized block (chapter 11).

## 9.5 The Paired *t*-Test

When the variable under study is measured is on an interval or ratio scale and is approximately normally distributed, the *t*-test for paired (matched) samples may be used. This test is equivalent to the one sample *t*-test discussed in chapter 8, since, in effect, we are dealing with one sample—the difference in the observed individuals before and after some treatment, or the difference between matched pairs of individuals under different specified conditions.

**Assumptions of the Test**

1. The variable is measured on an interval or ratio scale.
2. The distribution of the variable (differences) is approximately normal.
3. Each individual is measured twice, once before the specified treatment and again following the specified treatment; or matched pairs of individuals are measured.
4. The data constitute a random sample from the population of interest.

### • Example 9.3
A paired *t*-test

As an example of a before-and-after study, consider weight gain in an animal called a centim eater. The weights of 10 individuals were determined at one week of age and again at two weeks of age. We assume that their differences ($D$) follow a normal distribution. Since we expect their weights to increase over time, we use a one-tailed hypothesis test. The data are as follows.

| | Weight (in grams) at: | | |
|---|---|---|---|
| Animal Number | One week | Two weeks | $D$ |
| 1 | 1.0 | 2.5 | 1.5 |
| 2 | 1.3 | 2.6 | 1.3 |
| 3 | 0.9 | 2.5 | 1.6 |
| 4 | 0.8 | 2.2 | 1.4 |
| 5 | 1.1 | 2.6 | 1.5 |
| 6 | 1.2 | 1.0 | −0.2 |
| 7 | 0.7 | 2.0 | 1.3 |
| 8 | 0.6 | 2.1 | 1.5 |
| 9 | 1.5 | 2.6 | 1.1 |
| 10 | 2.0 | 3.6 | 1.6 |
| | | Mean $\bar{x}_D$ | 1.26 |
| | | Standard deviation = $S_D$ = | 0.536 |

The null hypothesis for this one-tailed test is that the population mean difference $(\mu_D)$ in weight between weeks 1 and 2 is less than or equal to zero, or

$$H_0\colon \mu_D \leq 0$$

and the alternative hypothesis is

$$H_a\colon \mu_D > 0$$

Recall that the critical value of $t$ is for a one-tailed test when alpha is set at 0.05 is found in table A.2 in the column headed "0.1."

We are interested in the difference in weight between one week and two weeks (the column headed "$D$") and the mean of these differences ($\bar{x}_D$), which is 1.26. This is the sample mean difference from week one to week two. The expression

$$t = \frac{\bar{x}_D}{s_{\bar{D}}} \tag{9.7}$$

has a $t$ distribution with $n - 1$ degrees of freedom, where $n$ is the number of matched pairs. The symbol $\bar{x}_D$ is the sample mean difference, and $s_{\bar{D}}$ is the standard error of the mean difference, calculated by

$$s_{\bar{D}} = \frac{s_D}{\sqrt{n}} \tag{9.8}$$

Note the distinction between the standard deviation ($S_D$) and the standard error ($s_{\bar{D}}$). For the example

$$s_{\bar{D}} = \frac{s}{\sqrt{n}} = \frac{0.536}{\sqrt{10}} = 0.169$$

and

$$t = \frac{\bar{x}_D}{s_{\bar{D}}} = \frac{1.26}{0.169} = 7.46$$

As before, we consult table A.2 to determine if $H_0$ may be rejected. For a one-sided test, the critical value of $t$ for 9 degrees of freedom at alpha = 0.05 is 1.833. The $p$ value, determined by consulting table A.2, is $p < 0.0001$. Therefore, we reject $H_0$ and conclude that there is a significant increase in weight between weeks 1 and 2. The weight change is statistically discernible. Whether the magnitude of the weight change (1.26) is important depends on our understanding of the biology of this animal.

If we wish, we could compute a confidence interval for the mean difference ($\mu_D$) by the same method used earlier for a single sample (section 7.5).

## 9.6 Nonparametric Tests for Two Related Samples

In matched pairs or repeated-measures experiments in which the data do not satisfy the assumptions of parametric tests, perhaps because the distribution of the variable is not approximately normal or because measurement is on an ordinal scale, one of two nonparametric tests for two related samples may be used.

### 9.6.1 The Sign Test

The sign test is a nonparametric test for paired or matched samples that is useful when the direction (< or >) of the difference between matched pairs can be determined but the magnitude of the difference cannot. The null hypothesis tested is that $p(A > B) = p(A < B)$, where A and B are measurements of the matched pair. In plain English this means that the probability that A is greater than B is equal to the probability that A is smaller than B for any given pair.

To conduct the test, each pair is assigned either a plus or a minus sign, depending on which member of the pair is larger. Tied pairs (no difference) are dropped from the analysis, and $n$ is reduced accordingly. This is essentially a binomial distribution problem in which we compare the frequency of pluses and minuses with the distribution expected under the null hypothesis:

$H_0$: frequency of pluses = frequency of minuses

with $p = 0.5$, $q = 0.5$, and $k$ = the number of paired observations ($n$). This is exactly analogous to determining the probability of obtaining $x$ heads and $k - x$ tails in $k$ tosses of a coin (section 5.1).

**• Example 9.4**

A football team was given a new sports beverage during summer training. Fifteen minutes after drinking it, each player was asked if he felt better than, worse than, or the same as before he drank the beverage. Nine reported feeling better, one reported feeling worse, and two reported no change. Using this sample of 12 players, we wish to know if drinking this beverage affects how players feel.

In this situation each individual is paired with himself or herself and is in effect measured twice—feeling before the beverage and feeling after the beverage. Thus, we can detect the direction of change in any individual (better or worse), but not the magnitude of the change. The sign test is appropriate for this situation and is conducted in the following manner.

Nine players reported feeling better and receive a plus sign, while one reported feeling worse and receives a minus sign. The two who reported no change are dropped from the sample, and the sample size is reduced accordingly. Let $x$ = the frequency of the less frequent sign (1, in this case) and $k - x$ = the frequency of the more frequent sign (9, in this case). The probability of obtaining 1 minus sign and 9 plus signs in 10 "trials" ($k$ = 10) is given by the now familiar expression (from section 5.1):

$$p(x) = \frac{k!}{x!(k-x)!} p^x q^{(k-x)}$$

As before, we are interested in the probability of the exact outcome we observed plus any more extreme outcomes (in this case 0 minuses and 10 pluses is the only more extreme outcome). For the example

$$p(1) = \frac{10!}{1! \times 9!} 0.5^1 \times 0.5^9 = 0.009760$$

and

$$p(0) = \frac{10!}{0! \times 10!} 0.5^0 \times 0.5^{10} = 0.000976$$

$$p(1) + p(0) = 0.009876 + 0.000976 = 0.010736$$

or a probability of approximately 0.011 of obtaining 1 minus value and 9 plus values or more extreme case if the null hypothesis is true. This is a one-tailed probability, but the way the example is phrased ("better" or "worse") indicates a two-tailed test. Thus, the probability of obtaining the observed result or one more extreme if $H_0$ is true is twice the one-tailed probability, or

0.010736 × 2 = 0.021472. Since the *p*-value is less than 0.05 (our usual choice for $\alpha$), we reject $H_0$ and conclude that drinking the beverage affected how the players felt.

Why is this a two-tailed test? We did not specify the direction of change in feeling in the null hypothesis. Had our research hypothesis been that taking the new sports drink makes players feel better, the null hypothesis would have been that taking the drink makes players feel no different or worse. That null hypothesis could be rejected with a *p* value of 0.010736 rather than one of 0.021472.

### 9.6.2 The Wilcoxon Signed-Ranks Test

This nonparametric test is appropriate when the direction of the difference between matched pairs can be determined and when the differences can be ranked with respect to each other. Thus, it is appropriate for data measured on an ordinal scale or for data measured on an interval or ratio scale when the variable is not normally distributed. Unlike the sign test, this test does give greater weight to pairs with larger differences than to pairs with smaller differences. The differences are ranked with respect to their absolute values (i.e., –1 has a lower value than either +2 or –2), but the sign of the difference is retained with the rank. As in the sign test, pairs in which there is no difference are dropped from the sample, and the sample size is reduced accordingly.

For this test, when $H_0$ is true, the sum of the positive ranks in a population should be about equal to the sum of the negative ranks in the population.

**• Example 9.5**

The Wilcoxon signed-ranks test

Female gully cats are thought to be more aggressive when they have kittens. Accordingly, aggressiveness scores on a scale of 1–10, with 10 being most aggressive, were obtained for a group of 7 females without kittens and for these same 7 females when they had kittens. The scores were as follows:

| Female ID # | Without Kittens | With Kittens | *D* | *R* |
|---|---|---|---|---|
| 1 | 3 | 7 | 4 | 4 |
| 2 | 2 | 8 | 6 | 6 |
| 3 | 5 | 4 | –1 | 1.5(–) |
| 4 | 6 | 9 | 3 | 3 |
| 5 | 5 | 10 | 5 | 5 |
| 6 | 1 | 9 | 8 | 7 |
| 7 | 8 | 9 | 1 | 1.5 |

Note that animals 3 and 7 were tied with a rank of 1.5. The fact that animal 3 had a difference in before-and-after scores of –1 and animal 7 had a difference of +1 makes no difference with respect to their ranks. However, the fact that animal 3 had a negative score is noted. To calculate the value of the test statistic, designated as T, the sum of ranks that have a negative sign and the sum of ranks that have a positive sign are calculated. T is the smaller of these 2 sums. For the example, T is 1.5. Critical values of T are given in table A.5. $H_0$ is rejected if the calculated value of T is equal to or *smaller* than the tabular value of T. In this case, the critical value is 2, so we reject the null hypothesis and conclude that gully cats are more aggressive when kittens are present.

Table A.5 gives values of *n* (sample size) up to 25. For larger samples, T is approximately normally distributed and the normal distribution may be used, where

$$z = \frac{T - \dfrac{n(n+1)}{4}}{\sqrt{\dfrac{n(n+1)(2n+1)}{24}}} \qquad (9.9)$$

$H_0$ is rejected if $|z|$ is equal to or larger than 1.96 for alpha 0.05.

## 9.7 Power of the Test: How Large a Sample Is Sufficient?

Consider the following situation. An industry releasing pollutant *X* into a stream would like to show government regulators that they are having no effect on concentrations of *X* in the stream. They already know that background levels of *X* are naturally quite variable. Their employees collect five water samples in stream locations upstream and five downstream from the discharge pipe and measure concentrations of *X* in each sample. Although the calculated sample mean downstream from the pipe is considerably greater than the mean upstream from the pipe, the variance for each group is large. The calculated *t*-statistic is small and thus $H_0$ (no difference) cannot be rejected. The industry reports to the regulator that there is no significant difference and thus they are having no clear effect on the stream. There is something fishy here, but what? Of course—the power of their test was small, making it exceedingly difficult to reject $H_0$ even when it's actually false!

In previous sections (8.4, 9.1) we have brought up the term "power." The **power of the test** is the chance of rejecting the null hypothesis when it is indeed false. Recall that power is equivalent to $1 - \beta$ (one minus the type II error rate). Although power is not yet commonly analyzed in biological statistics, researchers are encouraged to make wider use of power analysis.

In general, the power of a test is influenced by three properties that we can control:

1. The type I error rate ($\alpha$);
2. The difference between two means that we can discriminate ("**minimum detectable difference**", also called "**effect size**") ($\delta$);
3. Sample size ($n$)

In most cases, we set $\alpha$ to some fixed value, say 0.05. If we increase $\alpha$ to a higher error rate (say, 0.10), then $\beta$ declines and power goes up. Increasing is not always desirable. If the difference between the means is quite large, then we can more easily discriminate them. In other words, our $t$-test more likely reveals a significant difference when there is indeed a large difference between means. Finally, all things being equal, as sample size ($n$) goes up, power increases. This is because the sampling distributions for each of the means become narrower with increased sample size (section 7.2). The net effect is that, with larger samples, we are more likely to discriminate a real difference between groups. A simple graphic illustrates this effect (Fig. 9.1).

**Figure 9.1** An illustration of how the power of the test improves with increased sample size

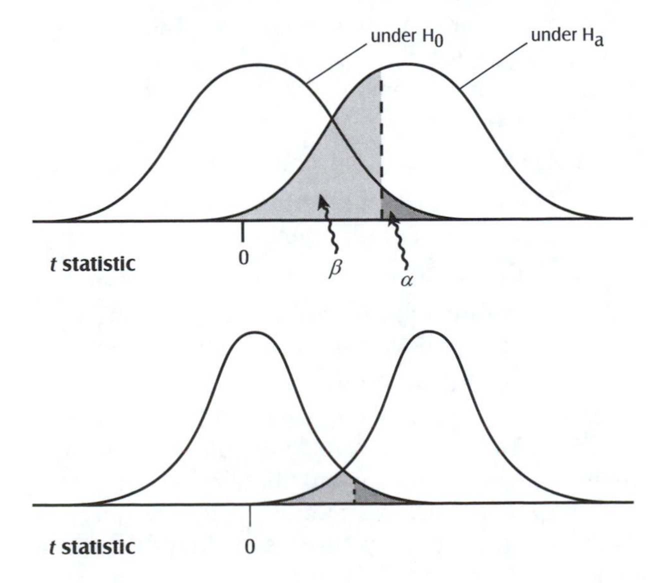


In figure 9.1, the distribution of the $t$ statistic is shown, both when $H_0$ is true and when it is false. When $H_0$ is true, a value of $t$ beyond a particular decision criterion (dashed vertical line) leads to a type I error with probability $\alpha$ (darker shaded area). When $H_0$ is false, the distribution of the $t$ statistic for a particular alternative hypothesis is also shown. Values of $t$ to the left of the decision criterion lead to accepting $H_0$ (a type II error), with probability $\beta$ (lighter shaded area). Power of the test is $1 - \beta$, the area under $H_a$ to the right of the decision criterion. When sample size is large, the distribution of the $t$ statistic is narrower (greater precision) and the areas of overlap between the two curves is less. Since we keep $\alpha$ constant (usually 0.05), $\beta$ is smaller and hence power is greater.

More formally, power analysis allows us to determine one of several properties of a sampling distribution. For instance, for a given level of $\alpha$ and sample size ($n$), we can determine power ($1 - \beta$) for a range of differences between the means of two groups. More commonly, we set power to some desired level and determine one of the two remaining properties.

### 9.7.1 Determining the Sample Size Needed to Detect a Minimum Effect

First, if we have some preliminary data, we can use power analysis to determine the sample size needed to detect a minimum difference between two groups. For example, we might want to be able to detect a blood-pressure change of at least 5mm Hg following a drug treatment. Preliminary data are needed to provide an estimate on the variance. By estimating the required sample sizes, we then have some guidance for designing an effective future experiment.

If we know the desired power and specify a minimum difference (effect size) between two groups ($\delta$), we can determine the sample size needed by equation 9.10. In order to perform this power analysis, we must have a prior estimate of the variance and assume that both populations are normally distributed. When both populations have the same variance (estimated by $s_p^2$), sample sizes ($n$) are equal for each group, and the degrees of freedom ($\gamma$) = 2(n − 1), the required sample size for a 2-sided $t$-test is:

$$n \approx \frac{2s_p^2}{\delta^2} \times (t_{\alpha,\gamma} + t_{2\beta,\gamma})^2 \tag{9.10}$$

An example should help illustrate.

---

• **Example 9.6**
Power analysis

Milk production (in kg/day) of dairy cattle was compared between two groups, those receiving a vitamin supplement and those not receiving the supplement. In a preliminary study, milk production was shown to be normally distributed with a variance of 0.64. For a 2-sided test with $\alpha = 0.05$, what sample size is necessary to detect a difference of at least 0.5 kg/day, while keeping the power at least 90%?

---

Power = $1 - \beta \geq 0.9$, so $\beta \leq 0.10$. We thus use $\beta = 0.10$ in our formula. Since the critical values for $t$ depend on degrees of freedom, which we don't know, we'll first choose the critical values from the last row of the table ($\infty$). $t_{0.05,\,\infty} = 1.960$ and $t_{0.20,\,\infty} = 1.282$. Now, substituting into the general formula:

$$n \approx \frac{2(0.64)}{(0.5)^2} \times (1.96 + 1.282)^2 = 53.8$$

In other words, at least 54 cows should be included in each group. If we refer back to table A.2, the critical values for $2(54 - 1) = 106$ degrees of freedom are close to the original numbers we picked, so the calculation should approximate our original results. $t_{0.05,\,106} = 1.99$ and $t_{0.20,\,106} = 1.29$. Now, substituting again:

$$n \approx \frac{2(0.64)}{(0.5)^2} \times (1.99 + 1.29)^2 = 55.1$$

To be safe, we should use at least 56 cows in each group in our next experiment.

### 9.7.2 Determining Minimum Detectable Difference

Suppose we are interested in how large a difference we could detect in a study that is already completed. If we know the sample size and variance and specify a particular power level, we can determine the minimum detectable difference (effect size, $\delta$) by equation 9.11.

$$\delta \geq \sqrt{\frac{2(s_p{}^2)}{n}} \times (t_{\alpha,\gamma} + t_{\beta,\gamma}) \qquad (9.11)$$

Continuing with our dairy-cattle example, imagine that we had obtained samples of size 30 from each group and we want to keep power of at least 90%. Substituting into this equation the appropriate critical values ($t_{0.05,29} = 2.045$, $t_{0.20,29} = 1.311$) and statistics ($s_p{}^2 = 0.64$, $n = 30$), we get:

$$\delta \geq \sqrt{\frac{2(0.64)}{30}} \times (2.045 + 1.311) =$$

$$\delta \geq (0.2066) \times (2.045 + 1.311) = 0.69 \text{ kg/day}$$

In other words, in this study of dairy cattle, the difference in milk yields between the vitamin-supplement and control groups would have to be at least 0.69 kg in order for us to detect a significant treatment effect.

In conclusion, power analysis can be a useful tool for experimental design, as it allows us to determine necessary sample sizes for future experiments. Power analysis also provides a method for showing how small an effect could have been detected by a particular experiment. This is particularly important when we find no significant difference between treatment groups. Power analysis allows others to see how sensitive our experiment was. In the case of the stream pollution described earlier, we would likely have found that the minimum detectable difference was quite large. A regulatory agency would do well to require a more sensitive study.

Some statistical reference books (e.g. Sokal and Rohlf 1995, Zar 1999) provide detailed graphics to aid in power determinations for various statistical tests. Some statistical software can also perform power analysis.

## 9.8 Review: Which Statistical Test Is Appropriate?

In the last two chapters, you have been introduced to a variety of tests for making inferences about means (or medians). Let's review the rationale for choosing an appropriate statistical test, which depends on answers to the following questions:

1. How many populations are we considering?
2. If more than one population, are the samples related or independent?
3. Which scale of measurement is used?
4. If the interval or ratio scale, are the data normally distributed? If not, are sample sizes large ($n > 30$)?

Recall that parametric statistics (like the $t$-test) require the interval or ratio scale of measurement and either a normal distribution or large sample sizes. For two sample comparisons, the following graphic may serve as a helpful key for choosing the correct test.

**Figure 9.2** A simple key to statistical tests for two-sample hypotheses

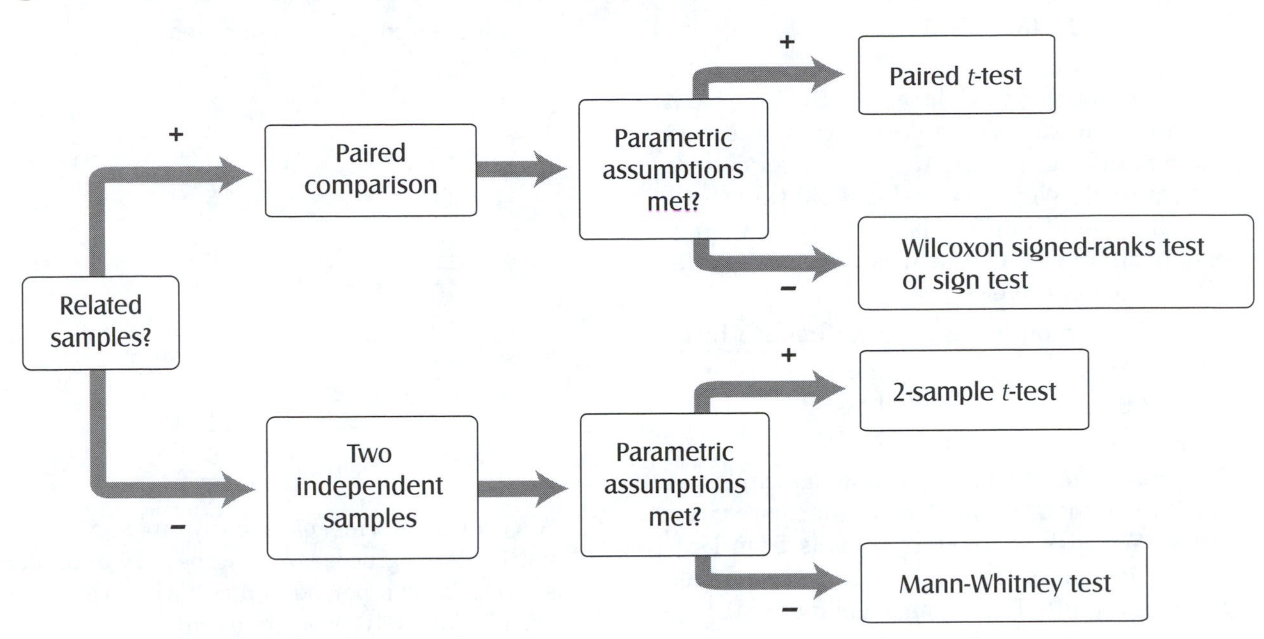


## 9.9 Comparisons of Variances from Two Samples

On occasion, we wish to know whether the variability within each of two populations is the same. For instance, a geneticist may wish to know whether the offspring generation is more variable for a trait than the parental generation. Or we may want to check assumptions of a statistical test (section 9.1). To compare two variances, we simply take the ratio of the two numbers (usually the larger over the smaller value) to generate an ***F* statistic**, which can then be compared to a critical value in an *F* table. Although we won't do this procedure here, we examine the *F* distribution in more detail in chapter 10.

## Key Terms

effect size
error variance
*F* statistic
independent samples
Mann-Whitney *U*-test
minimum detectable difference (= effect size)
paired comparison
power of the test
related samples
sign test
t-test
Wilcoxon signed-ranks test

## Exercises

For each of the following problems, state the appropriate null hypothesis and the alternative hypothesis, conduct the appropriate test, and state a conclusion. Reject $H_0$ if $p \le 0.05$. Be sure to include the *p* value. Assume the variable in question to be approximately normally distributed unless instructed otherwise.

### The *t*-Test for Independent Samples

9.1 Data on resting pulse rates (in bpm) were collected for random samples of 57 men and 63 women between the ages of 18 and 21. We wish to know if there is a difference in the mean pulse rate of men and women in this population. The results were as follows. Use a two-sample t-test to examine this hypothesis and also compute a 95% confidence interval for $\mu_m - \mu_f$. Be sure to interpret both. Are the results consistent? Explain.

| | Men | Women |
|---|---|---|
| $\bar{x}$ | 73.789 | 82.270 |
| $s$ | 10.395 | 13.750 |
| $n$ | 57 | 63 |

9.2 Data on reaction time (in milliseconds) for random samples of 58 men and 68 women were collected. We wish to know if there is a difference in reaction time between men and women in this population. The results were as follows. Use the same approach as in exercise 9.1.

| | Men | Women |
|---|---|---|
| $\overline{x}$ | 170.21 | 181.31 |
| $s$ | 32.643 | 45.988 |
| $n$ | 58 | 68 |

9.3 Random samples of largemouth bass and smallmouth bass were taken from a lake, and their standard lengths (in millimeters) were determined. We wish to know if the mean standard length differs between the two species in this lake. The results were as follows: (Data from J. Kagel.)

| | Largemouth Bass | Smallmouth Bass |
|---|---|---|
| $\overline{x}$ | 272.8 | 164.8 |
| $s$ | 96.4 | 40.0 |
| $n$ | 125 | 97 |

9.4 The mass (in grams) of random samples of adult male tuatara from two locations are given below. We wish to know if animals from location A have a larger mean mass than animals from location B. (Data from J. Gillingham).

| Location A | | Location B |
|---|---|---|
| 510 | 790 | 650 |
| 773 | 440 | 600 |
| 840 | 435 | 600 |
| 505 | 815 | 575 |
| 765 | 460 | 452 |
| 780 | 690 | 320 |
| 235 | | 660 |

9.5 Liver alcohol dehydrogenase activity in two random samples of catfish was determined. One sample was taken upstream from a brewery, and the other sample was taken downstream from the brewery. We wish to know if living downstream from a brewery increases liver alcohol dehydrogenase in these animals.

| Upstream | Downstream |
|---|---|
| 10 | 30 |
| 25 | 32 |
| 8 | 28 |
| 11 | 35 |
| 19 | 29 |
| 7 | 32 |
| 5 | 32 |
| 30 | 38 |
| | 31 |

9.6 The effect of copper sulfate on mucus cells in the gill filaments of a certain species of fish was investigated. We wish to know if exposure for 24 hours to copper sulfate reduces the number of mucus cells in this species. The number of mucus cells per square mm in the gill filaments of both untreated fish and exposed fish are below.

| Untreated | Exposed |
|---|---|
| 16 | 10 |
| 17 | 8 |
| 12 | 10 |
| 18 | 12 |
| 11 | 13 |
| 18 | 14 |
| 12 | 6 |
| 15 | 5 |
| 16 | 7 |
| 14 | 5 |
| 18 | 10 |
| 12 | 11 |
| | 9 |
| | 8 |

9.7 A certain species of bacterium was grown with either glucose or sucrose as a carbon source. After a period of incubation, the number of cells ($\times 10^6$) was determined. Is there a difference in growth rate of the bacterium between the two carbon sources?

| Glucose | Sucrose |
|---|---|
| 6.3 | 5.8 |
| 5.7 | 6.2 |
| 6.8 | 6.0 |
| 6.1 | 5.1 |
| 5.2 | 5.8 |

9.8 Six randomly selected pea plants were treated with a plant growth regulator, and six randomly selected plants were not treated. We wish to know if the growth regulator affects internode growth.

| Internode Length (in millimeters) in | |
|---|---|
| Treated Plants | Untreated Plants |
| 15.2 | 13.5 |
| 12.3 | 9.8 |
| 11.6 | 10.2 |
| 14.8 | 8.7 |
| 10.0 | 9.2 |
| 14.2 | 9.0 |

9.9 Growth of pine seedlings in a chemically defined liquid substrate with and without molybdenum was measured. We wish to know if growth is greater in the presence of molybdenum.

| Growth (centimeters/year) | |
|---|---|
| Without Molybdenum | With Molybdenum |
| 3.2 | 4.5 |
| 4.5 | 6.2 |
| 3.8 | 5.8 |
| 4.0 | 6.0 |
| 3.7 | 7.1 |
| 3.2 | 6.8 |
| 4.1 | 7.2 |

9.10 The surface pH of two lakes was measured at several randomly selected sites in each lake. We wish to know if the hydrogen-ion concentration of the two lakes is different. (Caution: pH is not a linear measurement! The data should be converted to hydrogen-ion concentrations.)

| Lake A | Lake B |
|---|---|
| 7.1 | 6.9 |
| 7.2 | 6.8 |
| 6.9 | 6.7 |
| 7.0 | 6.8 |
| 7.1 | 7.0 |

9.11 The soil pH in a coniferous forest and in a deciduous forest was measured at several randomly selected places in each site. We wish to know if the hydrogen-ion concentration is different in the two sites. (Caution: pH is not a linear measurement! The data should be converted to hydrogen-ion concentrations.)

| Coniferous | Deciduous |
|---|---|
| 5.9 | 6.2 |
| 6.0 | 6.4 |
| 6.2 | 6.1 |
| 5.8 | 6.3 |
| 5.6 | 6.4 |
| 5.7 | 6.0 |
| 5.8 | 6.2 |
| 5.7 | 5.9 |
| 5.6 | 6.1 |
| 5.9 | 6.0 |

9.12 Using the data in table B.3 (see digital appendix), determine if male athletes have faster reaction times than male nonathletes. (Caution: take care with interpreting what we mean by faster reaction time.)

9.13 Using the data in table B.3, determine if women smokers have faster pulse rates than women nonsmokers.

9.14 Using the data in table B.2 (see digital appendix), select random samples of 30 male mosquito fish and 30 female mosquito fish. From these samples determine if the population-mean length for females is greater than the population-mean length for males.

**The Mann-Whitney Test**

9.15 Do people find hairy spiders scarier than nonhairy spiders? To find out, 20 people were randomly assigned to two groups of 10 each. One group viewed a hairy spider, and the other group viewed a very similar but nonhairy spider. Each person was asked to rate the spider she or he viewed on a scariness scale from 1 to 10 (10 being most scary).

| Hairy | | Nonhairy | |
|---|---|---|---|
| 10 | 10 | 7 | 5 |
| 8 | 9 | 6 | 4 |
| 7 | 9 | 8 | 5 |
| 9 | 5 | 6 | 6 |
| 9 | 8 | 1 | 3 |

9.16 Male hoop snakes, upon encountering one another, may engage in a protracted ritualized combat behavior until one establishes himself as dominant over the other. We wish to know if these encounters last longer in the presence of a female. Twenty-four males were randomly assigned to pairs. Six randomly selected pairs were tested in the presence of a female, and six were tested in the absence of a female. This variable is probably not normally distributed. The results were as follows.

| Interaction Time (in minutes) | |
|---|---|
| Pairs without Female | Pairs with Female |
| 10 | 59 |
| 15 | 35 |
| 8 | 70 |
| 30 | 65 |
| 1 | 43 |
| 80 | 90 |

9.17 The 72-hour blastogenesis of chicken peripheral blood lymphocytes from a group treated with PHA and from an untreated group are given below. This variable is *not* approximately normally distributed. We wish to know if treatment with PHA has an effect on blastogenesis of these cells.

| Control | Treated |
|---|---|
| 1631 | 87700 |
| 50102 | 69553 |
| 1369 | 76215 |
| 41188 | 366 |
| 387 | 40104 |
| 498 | 38661 |
| 259 | 141153 |
| 329 | 154805 |
| 4330 | 123075 |
| 5002 | 627 |
| 658 | 126175 |
| 300 | 11223 |
| | 300 |

9.18 Seven tomato plants were treated with chlorogenic acid to determine if this influences the activity of the enzyme *o*-diphenol oxidase in their leaves. A control group of seven plants were not treated. We do not know if this variable is approximately normally distributed, nor is it possible to determine this with such a small sample. Does this treatment affect activity of the enzyme?

| Treated | Untreated |
|---|---|
| 35 | 10 |
| 45 | 18 |
| 36 | 8 |
| 11 | 29 |
| 41 | 17 |
| 29 | 8 |
| 38 | 11 |

9.19 Assume that in exercise 9.4 the variable is not approximately normally distributed. Conduct the appropriate test, and state your conclusions. Compare the results of this analysis with the results in exercise 9.4.

### The Paired *t*-Test

9.20 The wattle thickness (in millimeters) of 15 randomly selected chickens was measured before and after treatment with PHA. Does treatment with PHA increase wattle thickness? (Data from F. McCorkle.)

| Chicken Number | Pretreatment | Posttreatment |
|---|---|---|
| 1 | 1.05 | 3.48 |
| 2 | 1.01 | 5.02 |
| 3 | 0.78 | 5.37 |
| 4 | 0.98 | 5.45 |
| 5 | 0.81 | 5.37 |
| 6 | 0.95 | 3.92 |
| 7 | 1.00 | 6.54 |
| 8 | 0.83 | 3.42 |
| 9 | 0.78 | 3.72 |
| 10 | 1.05 | 3.25 |
| 11 | 1.04 | 3.66 |
| 12 | 1.03 | 3.12 |
| 13 | 0.95 | 4.22 |
| 14 | 1.46 | 2.53 |
| 15 | 0.78 | 4.39 |

9.21 Data on resting and post-exercise pulse rates were collected for 8 individuals between 19 and 22 years of age. We wish to know if there is a difference in preexercise and postexercise pulse rates.

| Individual Number | Resting | Postexercise |
|---|---|---|
| 1 | 108 | 136 |
| 2 | 60 | 90 |
| 3 | 70 | 78 |
| 4 | 54 | 108 |
| 5 | 54 | 102 |
| 6 | 72 | 92 |
| 7 | 101 | 118 |
| 8 | 96 | 176 |

9.22 For the situation in exercise 9.21, the following data were collected for preexercise and postexercise body temperature. Is there a difference in the two body temperatures?

| Individual Number | Resting | Postexercise |
|---|---|---|
| 1 | 99.0 | 99.4 |
| 2 | 97.8 | 98.1 |
| 3 | 98.6 | 98.6 |
| 4 | 98.7 | 98.7 |
| 5 | 98.7 | 98.7 |
| 6 | 98.2 | 98.2 |
| 7 | 98.7 | 98.8 |
| 8 | 98.6 | 99.2 |

9.23 Six laboratory mice were placed one at a time in a one-square-meter enclosure. The number of seconds in one minute that they were either near the wall or away from the wall was noted. We wish to know if mice spend more time by a wall than away from a wall (i.e., is there a "wall-seeking tendency" in this species?).

| Mouse Number | Seconds by Wall | Seconds away from Wall |
|---|---|---|
| 1 | 50 | 10 |
| 2 | 35 | 25 |
| 3 | 28 | 32 |
| 4 | 45 | 15 |
| 5 | 31 | 29 |
| 6 | 55 | 5 |

### The Wilcoxon Signed-Ranks Test

9.24 Ten individuals were asked to rate their feeling of well-being on a scale of 1 to 10 before and after taking an experimental drug. Does the drug change a person's sense of well-being?

| Individual Number | Before Drug | After Drug |
|---|---|---|
| 1 | 5 | 7 |
| 2 | 8 | 9 |
| 3 | 2 | 1 |
| 4 | 7 | 9 |
| 5 | 5 | 5 |
| 6 | 2 | 9 |
| 7 | 9 | 9 |
| 8 | 3 | 9 |
| 9 | 9 | 10 |
| 10 | 6 | 7 |

9.25 Assume that the variable in exercise 9.22 is not approximately normally distributed. Conduct a Wilcoxon test with these data.

### The Sign Test

9.26 Ten subjects were given an experimental drug and asked if their sense of well-being improved, became worse, or did not change after taking the drug. We wish to know if the drug is effective in improving one's sense of well-being.

7 reported an improvement
2 reported no change
1 reported feeling worse

9.27 Conduct a sign test using the data in exercise 9.22.

9.28 Conduct a sign test using the data in exercise 9.23.

### Choose the appropriate statistical test.

9.29 Aerial surveys of several randomly selected areas of forest land were used to determine damage by a certain insect. Some areas had been sprayed several years before the survey to control the insect and some had never been sprayed. A scale of 1 to 10 was used to assess damage, with 10 being most severe. We wish to know if there is a difference in previously sprayed areas and in areas that had never been sprayed.

| Sprayed Areas | Unsprayed Areas |
|---|---|
| 3 | 2 |
| 0 | 5 |
| 1 | 6 |
| 5 | 3 |
| 2 | 3 |
| 1 | 4 |
| 5 | 8 |
| 3 | 2 |
| 6 | 1 |
| 0 | 8 |
| | 2 |
| | 6 |
| | 5 |

9.30 Random samples of cranberries were collected from two bogs. We wish to know if the mean weight of cranberries (in grams) differs between the two bogs.

| | Bog A | Bog B |
|---|---|---|
| $\bar{x}$ | 1.31 | 1.27 |
| $s$ | 0.25 | 0.27 |
| $n$ | 25 | 27 |

9.31 *Brucella abortus* antibody titers (pfc/$10^6$ cells) in 15 turkeys were measured before and after a period of stress. We wish to know if stress affects antibody titer.

| Turkey Number | Before Stress | After Stress |
|---|---|---|
| 1 | 20 | 17 |
| 2 | 18 | 14 |
| 3 | 19 | 16 |
| 4 | 18 | 19 |
| 5 | 17 | 14 |
| 6 | 14 | 18 |
| 7 | 17 | 8 |
| 8 | 10 | 10 |
| 9 | 13 | 12 |
| 10 | 16 | 15 |
| 11 | 20 | 8 |
| 12 | 17 | 6 |
| 13 | 16 | 17 |
| 14 | 19 | 5 |
| 15 | 8 | 3 |

9.32 Assume that the variable in exercise 9.31 is not approximately normally distributed. Conduct the appropriate test with these data.

9.33 Spleen weights of road warblers infected with the avian malaria parasite and of those that were parasite-free were determined.

We wish to know if infection by this parasite affects spleen weight in this species.

| Infected | Healthy |
|---|---|
| 25.6 | 20.8 |
| 27.8 | 22.9 |
| 29.3 | 26.0 |
| 26.9 | 23.2 |
| 26.0 | 25.1 |
| 25.9 | 23.7 |
| | 25.6 |
| | 23.2 |

# 10

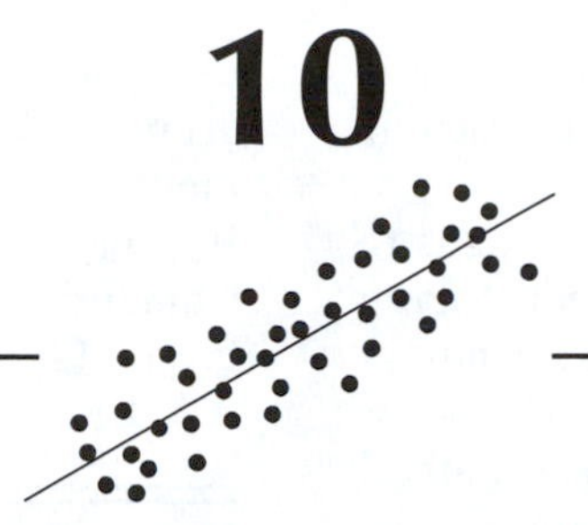

# Inferences Concerning Multiple Populations

## ANOVA

Analysis of variance (ANOVA) is one of the most versatile and useful techniques of statistical inference. It is used to analyze the effects of different groups (a categorical variable) on a measurement variable. Generally, we use ANOVA to answer the question: "Are the means from several groups all the same?" In this sense, ANOVA is an extension of the two-sample problem (chapter 9), although it uses a different approach. Essentially, ANOVA is a technique of partitioning the variance in a set of data into several components in such a way that the contribution of each of these components to the overall data set may be assessed. The techniques of ANOVA are useful both in situations where the researcher can carefully design and control experiments and for the analysis of certain types of observational (survey) data. One of the strengths of ANOVA is that it can be extended to a wide variety of experimental designs. In the next two chapters, we examine the basic concepts and most common applications of ANOVA. In the present chapter we explore the use of ANOVA to detect effects of groups of a single factor on the measurement (**one-way ANOVA**). We illustrate calculation steps for situations where sample sizes are equal within each group. With slight modifications, ANOVA can also be used with unequal sample sizes ("unbalanced data"), although space does not allow us to show examples of these cases. Since most investigators use computer software to do ANOVA, we include MINITAB output to compare with our worked examples.

## 10.1 The Rationale of ANOVA: An Illustration

A hypothetical example illustrates the rationale of analyzing the variance of samples when we are interested in their means. Fifteen juniper pythons, all of similar age and size, were randomly assigned to one of three groups. Group A received a certain drug; group B received another drug; and group C, the control group, received a placebo. The heart rate of each snake was then determined to see if either or both of the drugs affects heart rate. The null hypothesis in an experiment such as this is that the means of all three populations are equal, or

$$H_0: \mu_A = \mu_B = \mu_C$$

Population, in this case, refers to all possible juniper pythons that might be given a particular treatment.

One approach for solving a case like this is to conduct a series of two-sample *t*-tests, comparing each group with each other group (i.e., group A with group B, A with C, and B with C). However, it is not very efficient. Imagine that we instead had 10 different treatments we were comparing. Making all possible comparisons of pairs among 10 treatments would require 45 different *t*-tests! Running 45 tests would be quite time-consuming. Also, running so many tests increases the type I error rate. Suppose we set alpha at 0.05 for rejecting $H_0$ in any individual *t*-test (i.e., we will reject the null hypothesis that the means of any two groups are equal when $p \leq 0.05$). In such a situation there is a probability of 0.05 of making a type I error when comparing any two sample means. If we compared all possible group means for 10 groups (45 *t*-tests), the chance of a type I error would be far greater (about 0.90), causing us to incorrectly reject the null hypothesis that all of the group means are equal. Analysis of variance is designed to overcome this problem.

Let's return to our imaginary example with the three groups of juniper pythons. Because all of the individual juniper pythons in the study were chosen from the same group of animals, we may assume that all of the individuals involved in the experiment (the "experimental units") are more or less homogeneous. Any variation from one individual to another within a group is the normal variation we expect among individuals in a population. This variation is called **within-groups variance** or **error variance**. Because snakes from the available group of 15 were randomly assigned to the three treatment groups, we expect the error variance within each group to be approximately the same. Suppose we obtained the results shown in table 10.1

**Table 10.1** The effect of two drugs (A, B) and a placebo control (C) on the heart rate of juniper pythons (in beats per minute)

| Group A | Group B | Group C |
|---|---|---|
| 13 | 7 | 3 |
| 15 | 6 | 5 |
| 12 | 10 | 2 |
| 14 | 9 | 4 |
| 11 | 8 | 1 |
| $\bar{X} = 13$ | $\bar{X} = 8$ | $\bar{X} = 3$ |
| $s^2 = 2.50$ | $s^2 = 2.50$ | $s^2 = 2.50$ |
| | $s_t^2 = 19.124$ | |

Here the error variance is the same for each group (2.50). Now imagine that all 15 measurements were lumped together (figure 10.1). Because some measurements are quite different from others, this variance ($s_t^2$) is considerably larger. (Try making the calculation to verify.)

**Figure 10.1** Thought experiment using the juniper python heart-rate data. The within-group variance is small for each group (top). It is much larger in a single group (bottom)

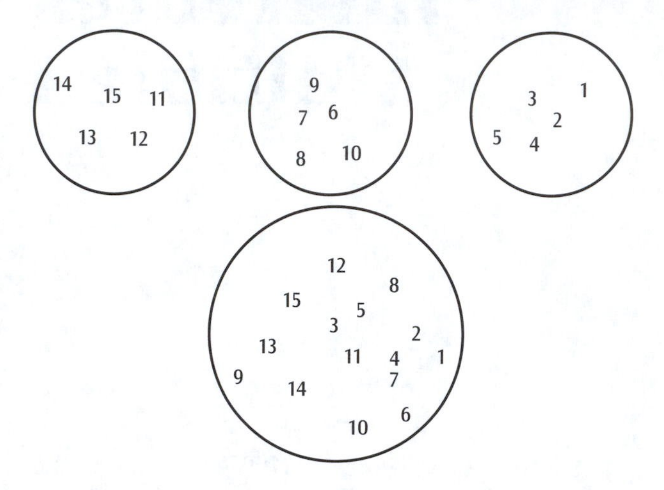


The approach with ANOVA is to partition the variance in a set of data into various components and to determine the contribution of each of these components to the overall variation. First, we consider the overall variation in the complete data set (all 15 snakes considered together). This is called the total variance, and it is about 19.124 in this case. There are two sources of this variation. One source is the usual variation from one individual to another caused by genetic and environmental differences among individuals; it has nothing to do with the treatments the animals were given. This is called the within-groups variance (error variance). Note in table 10.1 that the error variance in each group is the same (about 2.5). There is quite a difference between this error variance and the total variance, which tells us there is a source of variance that is not accounted for by the variation from one individual to another. Its source is the variance introduced by our treatments. The difference between the total variance in the data set and the within-groups variance is called the **among-groups variance** (also called **treatment variance**).

The sampling distribution that describes the ratio of these two variances, or

$$\frac{\text{among-groups variance}}{\text{within-groups variance}}$$

is the ***F* distribution**. The *F* distribution is a probability distribution, the shape of which depends on two types of degrees of freedom (*df*), those associated with the numerator and those associated with the denominator of the above expression. The among-groups degrees of freedom is determined as the number of treatments minus one and the within-groups degrees of freedom is the total number of observations in all groups minus the number of groups. A typical *F* distribution is shown in figure 10.2. The shaded portion of the curve is 0.05 of the total, and it is termed alpha, as before. The values of *F* that delimit this alpha region are found in table A.6, whose use will be described later.

**Figure 10.2** An *F* distribution. The shaded area illustrates alpha

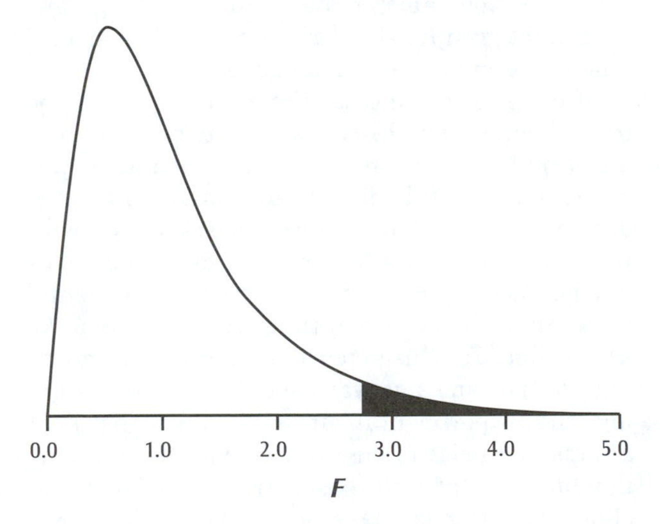


When treatments have no effect, virtually all of the variance in the data set is due to within-groups variance, which, you will recall, is the normal variance among individuals in a population. In this case, the variance ratio will be low (generally close to 1.0). When treatments have a real effect, the variance ratio will be larger (i.e., $F >> 1.0$).

If, at this point, you find yourself confused, don't be concerned. Confusion, they say, is the beginning of learning! Consider another outcome of this same imaginary experiment, shown in table 10.2. In this case the error variance and the total variance are practically the same, and we would probably conclude that most of the variation within the data set is attributable to the usual variation between individuals and that little or none of it results from the treatments given. In this case we would no doubt accept $H_0$ and conclude that neither drug has an effect on heart rate. How we reach this decision is explained in the following sections.

**Table 10.2** The effect of two drugs and a control on the heart rate of juniper pythons (in beats per minute)

| | Group A | Group B | Group C |
|---|---|---|---|
| | 1 | 3 | 2 |
| | 3 | 5 | 1 |
| | 2 | 1 | 5 |
| | 5 | 4 | 4 |
| | 4 | 2 | 3 |
| $\overline{X}$ | 3 | 3 | 3 |
| $s^2$ | 2.50 | 2.50 | 2.50 |
| | | $s_t^2 = 2.14$ | |

## 10.2 The Assumptions of ANOVA

There are several important assumptions of ANOVA, which should be adhered to rather closely. When these assumptions are not met, we cannot trust the results of the analysis. (Theorists tell us that alpha is not what we believe it to be in those cases.)

**General Assumptions of ANOVA**

1. Each of the groups is a random sample from the population of interest.
2. The measured variable is continuous (or if discrete, it may assume a large range of values).
3. Measurement is on a ratio or interval scale.
4. The error variances are equal.
5. The variable is approximately normally distributed.

Assumption 1 may not be violated! Randomization is usually achieved by randomly assigning the available experimental subjects to the various treatment groups, as in the juniper python example. If assumption 1 does not hold, we might as well throw out the data. When any of assumptions 2 through 5 are not met, one should consider use of a nonparametric method, discussed later in this chapter (section 10.5.2). The assumption that the error variances are equal should always be checked. We explain this procedure later in this chapter (section 10.4).

A variety of experimental designs can be analyzed with ANOVA, and each of these requires a somewhat different set of computations. Which design is chosen depends on the nature of the question being asked. In the following sections, we examine different forms of one-way ANOVA and its nonparametric analogue, the Kruskal-Wallis test. Other designs are introduced in chapter 11.

## 10.3 Fixed-Effects ANOVA (Model I)

Example 10.1 is an imaginary but fairly typical situation in which fixed-effects ANOVA is useful. There are three sample means representing three treatments, one of which serves as the control. We wish to know if either or both of the treatments are significantly different from the control. These are random samples because the individual subjects (experimental units) were randomly assigned to the three groups.

### • Example 10.1

An experiment using the fixed-effects design

A group of 30 highly inbred mice, just weaned, were randomly divided into three groups of 10 mice each and were given three different diets. At the end of several weeks, the gain in weight of each mouse was determined. Group 1 (the control group) was fed regular laboratory mouse food; group 2 was fed potato chips, Twinkies, and diet cola (the junk food diet); and group 3 was fed granola and organically grown prune juice (the health food diet). The results of our imaginary experiment are shown below. We wish to determine if the mean weight gain in any of the three treatments (groups) is significantly different from the other means.

**Weight Gain in Mice**

| Group 1 (Control) | Group 2 (Junk Food) | Group 3 (Health Food) |
|---|---|---|
| 10.8 | 12.7 | 9.8 |
| 11.0 | 13.9 | 8.6 |
| 9.7 | 11.8 | 8.0 |
| 10.1 | 13.0 | 7.5 |
| 11.2 | 11.0 | 9.0 |
| 9.8 | 10.9 | 10.0 |
| 10.5 | 13.6 | 8.1 |
| 9.5 | 10.9 | 7.8 |
| 10.0 | 11.5 | 7.9 |
| 10.2 | 12.8 | 9.1 |
| $\bar{x}_1 = 10.28$ | $\bar{x}_2 = 12.21$ | $\bar{x}_3 = 8.58$ |
| $n_1 = 10$ | $n_2 = 10$ | $n_3 = 10$ |

Inspection of the sample means above suggests that the population means are not all the same. However, a statistical analysis using ANOVA is necessary to verify that our intuition is correct. The null hypothesis for this analysis is:

$$H_0: \mu_1 = \mu_2 = \mu_3$$

and the alternative hypothesis is:

$$H_a: \text{not all } \mu\text{'s are equal.}$$

An experiment of this type is referred to as a **fixed-effects ANOVA (model I)**. "Fixed effects" means that the treatments used in the experiment are chosen by the investigator. This design may also be called a one-way ANOVA with fixed effects, since each observation in the data set is classified according to only one criterion—the group to which it is assigned (in this case, diet). The experimental units (individual mice) are randomly assigned to the various treatment groups by the investigator. Since each experimental unit is from the same population, it is assumed that the variation among individuals within any treatment group is the variation usually expected among individuals in a population.

Before beginning a discussion of how one tests the null hypothesis, we need to consider how and why this experiment was designed as it was. First, it is helpful in such an experiment if the error variance—the variance not associated with the treatments—can be kept to a minimum. This often is accomplished by selecting experimental units (mice, in this case) that are as nearly alike as possible. For this experiment a group of inbred mice of the same age was used. Randomly assigning the experimental units to the treatment groups is crucial to ensure that variation among individuals that still exists (in spite of our best efforts to minimize it) is not associated with one group more than with any other, and that the variance within each group (the error variance) will be equal among all the groups.

Once the sample has been properly selected and the experiment begins, it is necessary to ensure that all groups are treated identically, as far as possible, with respect to housing, water, temperature, and every other way except for the factor being tested (in this case, diet). Only then may we feel confident that any difference in weight gain noted among the three groups is caused by diet and not by some other factor.

### 10.3.1 Testing the Null Hypothesis That All Treatment Means Are Equal

We use ANOVA to test the null hypothesis that all of the treatment means are equal, or

$$H_0: \mu_1 = \mu_2 = \ldots = \mu_a$$

Recall that we expect the ratio of the among-groups variance to the within-groups variance (the variance ratio, or $F$) to be approximately 1.0 when the null hypothesis is true, and that the sampling distribution that describes the variance ratio is the $F$ distribution.

Example 10.1 further illustrates the calculations for obtaining $F$ where sample sizes are equal within each group. The data from this experiment are repeated and preliminary calculations needed to compute the variance ratio are given in table 10.3. We will describe these calculations step by step.

These calculations are used to determine the various sums of squares, which are later used to determine variances. The variances, in turn, are used to determine the variance ratio ($F$), discussed above. Note that the steps above have numbers corresponding to the descriptions below.

**Tips for the wise:** These calculations can be a bit tedious, so some tricks will save you much time and anxiety! First, note that the layout of the table above resembles a spreadsheet. A computer spreadsheet is the simplest way to do the calculations, since many of them involve computing sums. If you use a calculator, make liberal use of its memory, since the sums can be easily recalled. Finally, as a check on your answers, all variance terms you compute should be positive numbers.

1. Sum the observations for each group and then sum across all groups to get a **grand total**. The grand total may also be coded as $\Sigma\Sigma x$. In the current example, $\Sigma\Sigma x = 310.7$.
2. Sum the squared observations for each group and sum across all groups. In the current example, this total, $\Sigma\Sigma x^2 = 3305.09$. Note the difference between this set of operations and the operation in step 4 below.
3. Square each of the group totals, divide by $n$, and then sum across all groups. For instance, in the current example for the first group, 102.8 squared equals 10,567.84, which divided by the sample size for the group ($n = 10$) equals 1,056.784. Summing across all three groups gives 3283.789.
4. Square the grand total and divide by the total number of observations from all groups. This result is sometimes called the **correction term**. In the current example, where sample sizes within each group are equal, the total number of observations ($a \times n$) = 3 × 10 = 30. So, 310.7 squared and

**Table 10.3** Preliminary calculations for weight gain of mice fed different diets

| | Weight Gain in Mice | | | | | |
|---|---|---|---|---|---|---|
| | Group 1 (Control) | Group 2 (Junk Food) | Group 3 (Health Food) | | | |
| | 10.8 | 12.7 | 9.8 | | | |
| | 11.0 | 13.9 | 8.6 | | | |
| | 9.7 | 11.8 | 8.0 | | | |
| | 10.1 | 13.0 | 7.5 | | | |
| | 11.2 | 11.0 | 9.0 | | | |
| | 9.8 | 10.9 | 10.0 | | | |
| | 10.5 | 13.6 | 8.1 | | | |
| | 9.5 | 10.9 | 7.8 | | | |
| | 10.0 | 11.5 | 7.9 | | | |
| | 10.2 | 12.8 | 9.1 | | | |
| Group sums | | | | | Grand totals | Step |
| $\Sigma x$ | 102.8 | 122.1 | 85.8 | | **310.7** | **1** |
| $\Sigma x^2$ | 1059.76 | 1502.41 | 742.92 | | **3305.09** | **2** |
| $(\Sigma x)^2/n$ | 1056.784 | 1490.841 | 736.164 | | **3283.789** | **3** |
| | | | | $(\Sigma\Sigma x)^2/an$ | **3217.816** | **4** |
| | | | | SS total | **87.274** | **5** |
| | | | | SS among | **65.973** | **6** |
| | | | | SS within | **21.301** | **7** |

Data from example 10.1. In this example, the number of groups ($a$) is 3 and the sample size within each group ($n$) is 10.

then divided by 30 is 3217.816333 or about 3217.816.

5. Now we are ready to subtract to obtain the various sums of squares. To determine the **total sum of squares**, subtract the correction term from the sum of squared observations. In other words, **step 2 – step 4**, which in this case would be 3305.09 – 3217.816 = 87.274.
6. Determine the **sum of squares among groups** by subtracting the correction term from the answer to step 3. In other words, **step 3 – step 4**, which in this case would be 3283.789 – 3217.816 = 65.973.
7. Finally, determine the **sum of squares within groups** by subtracting the sum of squares among groups from the total sum of squares. In other words, **step 5 – step 6**, which in this case would be 87.274 – 65.973 = 21.301.

To summarize, we obtain the sums of squares as follows:

$$SS_{Total} = \sum^{a}\sum^{n} x^2 - \frac{(\sum\sum x)^2}{an} \quad (10.1)$$

$$SS_{Among} = \sum^{a} \frac{(\sum^{n} x)^2}{n} - \frac{(\sum\sum x)^2}{an} \quad (10.2)$$

$$SS_{Within} = SS_{Total} - SS_{Among} \quad (10.3)$$

We next can determine the different variance terms. In ANOVA, the variance is also called the **mean square** (MS). To convert these sums of squares into mean squares, we must divide each by their appropriate degrees of freedom. The among-groups **degrees of freedom** ($df_a$) is the number of groups minus one ($a$ – 1). The total degrees of freedom ($df_t$) is $an$ – 1, and the within-group degrees of freedom ($df_w$) is obtained by subtracting the among-groups degrees of freedom from the total degrees of freedom. We usually arrange the sums of squares, degrees of freedom, and mean squares in tabular form like tables 10.4 and 10.5. The mean squares of interest are obtained by dividing the sums of squares by their appropriate degrees of freedom. $F$ (the variance ratio) is obtained by dividing the among-groups mean square by the within-group mean square.

Work through the example to ensure comprehension! We now consult table A.6, "Critical Values of the $F$ Distribution," to see if we may reject $H_0$. The top row of the table gives degrees of freedom for the numerator mean square (the among-groups mean square), and the left column gives degrees of freedom for the denominator mean square (the within-groups mean square). In this case the degrees of freedom are 2 and 27. If our calculated value of $F$ is equal to or greater than the tabular value of $F$, we may reject $H_0$. The table value for 2 and 27 degrees of freedom (2 and 25, since 2 and 27 is not tabulated) is 3.39. Since our calculated value is much greater than this, we may reject $H_0$ and conclude that the means of the three treatments are not all equal, $p < 0.05$ (i.e., that one or more of the treatment means differs from the others).

**Table 10.4** Generalized ANOVA table

| Source of Variation | SS | df | MS | F |
|---|---|---|---|---|
| Among-groups | $SS_a$ | $df_a$ | $\frac{SS_a}{df_a}$ | $\frac{MS_a}{MS_w}$ |
| Within-groups | $SS_w$ | $df_w$ | $\frac{SS_w}{df_w}$ | |
| Total | $SS_t$ | $df_t$ | | |

**Table 10.5** ANOVA table for mouse diet example

| Source of Variation | SS | df | MS | F |
|---|---|---|---|---|
| Among-groups | 65.973 | 2 | 32.98 | 41.81 |
| Within-groups | 21.301 | 27 | 0.789 | |
| Total | 87.274 | 29 | | |

Computer software can run an ANOVA much faster than these hand calculations. If you have software available, run the analysis on this example and confirm that you obtain the same answer in the ANOVA table. An example from MINITAB appears below. Note in particular the ANOVA table, as well as summary statistics and plots of confidence intervals below the table. We will discuss the Tukey test in the next section.

### 10.3.2 Multiple Comparisons

Which means are different from which other means? At this point we cannot say. However, inspection of the confidence intervals computed by MINITAB suggest that the mean from the junk food group is larger than the other two groups. There are a number of different techniques for testing the differences between individual means following an ANOVA. These test procedures are called **multiple comparisons** tests. The specific technique depends to some extent on whether the investigator planned to compare certain groups of means or individual means with certain other

**Table 10.6** MINITAB results for example 10.1

**One-way ANOVA: WeightGain versus Group**

```
Source   DF       SS       MS       F      P
Group     2   65.973   32.986   41.81  0.000
Error    27   21.301    0.789
Total    29   87.274

S = 0.8882     R-Sq = 75.59%     R-Sq(adj) = 73.78%

                                        Individual 95% CIs For Mean Based on
                                        Pooled StDev
Level       N     Mean    StDev   ---+---------+---------+---------+------
Control    10   10.280    0.575              (----*---)
HealthFood 10    8.580    0.866   (----*---)
JunkFood   10   12.210    1.134                          (----*----)
                                  ---+---------+---------+---------+------
                                    8.4       9.6      10.8      12.0

Pooled StDev = 0.888

Tukey 95% Simultaneous Confidence Intervals
All Pairwise Comparisons among Levels of Group

Individual confidence level = 98.04%

Group = Control subtracted from:
Group         Lower    Center     Upper
HealthFood  -2.6859   -1.7000   -0.7141
JunkFood     0.9441    1.9300    2.9159

Group = HealthFood subtracted from:
Group         Lower    Center     Upper
JunkFood     2.6441    3.6300    4.6159
```

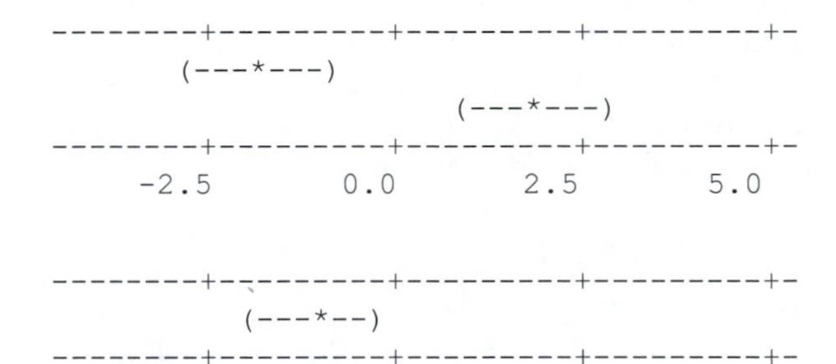


means before the experiment was conducted. These are called **planned comparisons**. In other cases the investigator does not know before the experiment is conducted which means are to be compared to which other means, and in fact, may wish to compare all possible pairs of means with each other. These are called **unplanned comparisons**. In this chapter we will deal with only one technique for unplanned comparisons, the extended **Tukey test** for multiple comparisons, because it is computationally relatively simple. There are a number of other such tests that one may use. Detailed descriptions are given in Sokal and Rohlf (1995). These procedures are typically listed by computer statistics packages as options in their ANOVA routines.

In example 10.1 we rejected $H_0$: $\mu_A = \mu_B = \mu_C$. The error mean square for this test was 0.789 with 27 df, and $n$ for all three treatments was 10. We may now compute a critical value that the difference between any two means must equal or exceed to be considered significantly different from each other. This critical value (CV) for samples of equal size is given by

$$CV = q\left(\sqrt{\frac{MS_e}{n}}\right) \tag{10.4}$$

where $MS_e$ is the error mean square, $n$ is the sample size (within each group), and $q$ is the studentized range value, found in table A.7, "Critical Values of $q$ for the Tukey Test." The top row of the table gives the number of groups, designated by $a$, and the left column gives the error degrees of freedom. The cell entry is the value of $q$. Since 27 df is not listed, we go to the next lower value listed (at 24 df), which is 3.53. Any error caused by doing this is a conservative one. (If a more accurate value for $q$ is needed, it may be obtained by interpolation.) Thus, the critical value is

$$CV = 3.53\left(\sqrt{\frac{0.789}{10}}\right) = 0.9915$$

We now calculate the absolute difference between each pair of means.

Control – Junk food: | 10.28 – 12.21 | = 1.93
Control – Health food: | 10.28 – 8.58 | = 1.70
Junk food – Health food: | 12.25 – 8.58 | = 3.67

Since the differences between the various pairs of means are all larger than the CV, we may conclude that all of the means are significantly different from each other at the 0.05 level of significance. A diet of potato chips, Twinkies, and diet cola induces a greater rate of weight gain in laboratory mice than does regular mouse food, which in turn induces more weight gain than does granola and organically grown prune juice.

An alternative way of considering this same information is to compute confidence intervals for the difference between each pair of means (recall section 9.2), making a slight modification of the critical value. This is displayed by MINITAB in table 10.6. The fact that all the confidence intervals fail to include zero indicates that each of the means is different from one another.

### 10.3.3 Fixed-Effects ANOVA Using Survey Data

"Treatments" in ANOVA are not necessarily something that an experimenter "does" to groups of individuals, as in example 10.1 (a **manipulative experiment**). Rather, they may be some factor that differs from population to population, but only if we feel relatively confident that the populations are similar except with respect to this factor (this is not always a valid assumption). The following example may help to clarify this.

#### Example 10.2

A fixed-effects ANOVA design using survey data

Random samples of 20 sticklebacks each were collected from three small lakes and three small streams in a given area. We wish to know if these populations differ in total length (in millimeters) among these six habitats. The data are shown in table 10.7. (Data are also available in digital appendix 4.)

**Table 10.7** Data for example 10.2 Body length (mm) of sticklebacks

| Lake A | Lake B | Lake C | Stream A | Stream B | Stream C |
|---|---|---|---|---|---|
| 31 | 36 | 28 | 47 | 47 | 38 |
| 32 | 30 | 38 | 48 | 37 | 36 |
| 34 | 32 | 31 | 50 | 41 | 48 |
| 34 | 37 | 32 | 42 | 38 | 43 |
| 35 | 35 | 29 | 44 | 32 | 42 |
| 30 | 32 | 38 | 34 | 45 | 31 |
| 33 | 32 | 40 | 41 | 42 | 40 |
| 32 | 37 | 36 | 40 | 40 | 45 |
| 37 | 39 | 43 | 44 | 43 | 42 |
| 33 | 28 | 34 | 47 | 40 | 49 |
| 36 | 32 | 32 | 39 | 39 | 39 |
| 30 | 31 | 39 | 47 | 45 | 30 |
| 32 | 35 | 31 | 43 | 41 | 42 |
| 39 | 40 | 36 | 40 | 39 | 39 |
| 30 | 36 | 28 | 38 | 32 | 38 |
| 29 | 31 | 39 | 32 | 48 | 35 |
| 42 | 32 | 32 | 41 | 32 | 49 |
| 39 | 27 | 38 | 45 | 45 | 40 |
| 37 | 35 | 29 | 42 | 41 | 43 |
| 29 | 31 | 32 | 37 | 38 | 42 |

This is a fixed-effects model, since the habitats (treatments) were not chosen at random but were selected by the investigator. The term "treatment" in this situation does not have its usual meaning, since nothing was treated. Rather, treatment refers here to the locations from which the fish were sampled. Note that in this situation the investigator did not manipulate the situation beyond choosing the habitats to be sampled. Thus, we are using survey data rather than data from a manipulative experiment.

The calculations for a fixed-effects ANOVA using observational data are exactly the same as when using experimental data. The major difference is in the interpretation. In example 10.1 we can be fairly certain that different diets caused differences in weight gain in our experimental animals because each animal was randomly assigned to each treatment group. In situations like example 10.2, it would not be accurate to conclude that different lakes or streams cause differences in the size of sticklebacks (if such differences are revealed by the ANOVA) because individual fish were not randomly assigned to the various habitats; they were already there. Thus, while we may detect differences among habitats, we should not conclude that the habitats directly cause the differences in the same sense that the diets in example 10.1 could be considered to be the cause of the differences in weight gain. If we detect differences that correspond to particular environmental features (such as, say, temperature), we could say that body length was associated with this feature. But a manipulative experiment is required to verify its cause.

If you worked your way through example 10.1 and tried your hand at a couple of the problems at the end of this chapter, you have developed a feeling about ANOVA by calculator, pencil, and paper. If you are like most people, you found the process tedious! Try example 10.2 using computer software. Table 10.8 is that analysis done with MINITAB.

The analysis clearly shows that the sticklebacks differed in mean body length among the six populations. The non-overlapping confidence intervals between the lake and stream populations suggest each habitat type forms a distinctive group. A follow-up Tukey test confirmed that these differences are real.

Following a multiple-comparisons test, most users summarize the results with a graphic. A simple approach is to list the groups (or their codes) in order of increasing sample means and draw lines connecting groups that are not significantly different from one another. For example 10.2, we might use the following summary graphic, taking the means listed in table 10.8:

LB LA LC   SB SC SA

### 10.3.4 One-Way ANOVA Design with Random Effects

In examples 10.1 and 10.2 we dealt with fixed-effects models (sometimes called model I ANOVA) because the treatments (diets, habitats) were selected by the investigator and we had no interest in generalizing the results to other diets or other habitats. In other words the questions asked were "Do these three diets differ with respect to weight gain in mice?" and "Is there a difference in the size of sticklebacks among these three lakes and three streams?"

On occasion we wish to ask a somewhat different question, such as "Is diet an important factor in weight gain?" In such a case, we might select a number of different diets at *random* from a large population of different possible diets and test these much as we did earlier. In this case, however, the treatments (diets) are not selected by the investigator but rather are chosen at random from a *population* of possible diets (thus, random effects). An ANOVA design using random effects is sometimes called a **random-effects (model II) ANOVA**. In this case the null hypothesis is that all diets are equal, and the alternative hypothesis is that one or more of the diets are different from the others.

We do not conduct a multiple-comparisons test in this type of ANOVA if the null hypothesis is rejected, since we have no interest in which diets are different from which others. Why? If the experiment were repeated, it is highly likely that different treatments (which, as you recall,

**Table 10.8** MINITAB output for one-way ANOVA on stickleback data (from table 10.7)

**One-way ANOVA: BodyLength versus Location**

```
Source     DF        SS       MS       F       P
Location    5    1585.5    317.1   16.38   0.000
Error     114    2206.4     19.4
Total     119    3791.9

S = 4.399     R-Sq = 41.81%     R-Sq(adj) = 39.26%

                                  Individual 95% CIs For Mean Based on
                                  Pooled StDev
Level      N    Mean   StDev   +---------+---------+---------+---------
Lake A    20  33.700   3.672    (----*-----)
Lake B    20  33.400   3.485   (----*-----)
Lake C    20  34.250   4.459     (-----*----)
Stream A  20  42.050   4.685                         (----*-----)
Stream B  20  40.250   4.667                     (-----*-----)
Stream C  20  40.550   5.186                      (-----*----)
                               +---------+---------+---------+---------
                            31.5      35.0      38.5      42.0

Pooled StDev = 4.399
```

were selected at random) would be used. Our question was "Does diet affect weight gain in mice?" If the overall ANOVA is significant, then the answer is yes.

Let's consider another example. A fisheries biologist wishes to determine if the fecundity of a particular species of fish might be influenced by habitat (i.e., if different lakes differed with respect to the fecundity of this species). To answer this question, the biologist might randomly select a number of lakes from a larger population of lakes and conduct a one-way ANOVA using the random-effects model. The ANOVA is conducted exactly as in the previous examples, except that multiple comparisons are not conducted. If the null hypothesis is rejected, the conclusion is that habitat affects fecundity. How it might differ from one particular lake to another is not a question of interest in the random-effects model, since if another random sample had been chosen, it would very likely include different lakes. We might want to know how much of the variation in fecundity is due to differences *among* populations versus differences among individuals *within* populations. This basically examines a breakdown of variance components. Such questions are very common in the field of quantitative genetics. Exploring this analysis is beyond the scope of this text, so interested readers should consult another text, such as Sokal and Rohlf (1995), for a complete treatment.

## 10.4 Testing the Assumptions of ANOVA

As we have seen (section 10.2), for ANOVA to work correctly, several assumptions must be met. A proper experimental design should assure that subjects are selected at random, which also assures independence of errors. The errors are "unexplained noise" that is not accounted by the treatments we impose. These errors are sometimes called **residuals**, the deviation of a measurement from its group mean. Recall that we earlier estimated the within-group variance, also known as the error variance. This is simply the variance of these residuals.

Assuming that the treatments do not in some way affect variability of a measured response, when individuals from a population are randomly assigned to treatments in a carefully controlled experiment, homogeneity of within-groups variance usually results. In example 10.1, which is a fairly typical application of a completely randomized design, the use of genetically similar individuals (inbred mice) randomized among treatment groups should result in homogeneity of error variances. Consider example 10.2, concerning the length of sticklebacks from six different habitats. In this case individuals could not be randomly assigned to the different habitats, since they were already there, and it would be unwise to assume that the variances among these six populations are equal. Descriptive statistics from this example are presented in table 10.9.

**Table 10.9** Group means and variances for sticklebacks from example 10.2

| | $\bar{X}$ | $s^2$ |
|---|---|---|
| Lake A | 33.70 | 13.484 |
| Lake B | 33.40 | 12.147 |
| Lake C | 34.25 | 19.882 |
| Stream A | 42.05 | 21.945 |
| Stream B | 40.25 | 21.776 |
| Stream C | 40.55 | 26.892 |

Typically, when using ANOVA, we check two assumptions, normality (of residuals) and homogeneity of (error) variance. In chapter 6, we see how normality can be checked using a graphic technique. Here we briefly see how these assumptions are checked using statistical software. Most software packages that do ANOVA have routines for checking these assumptions. For example, using MINITAB, we checked these assumptions for the stickleback data and display the results in figures 10.3 and 10.4.

On the normality plot, note that the points fall close to a straight line, indicating no significant departure from normality. This result was further confirmed by a goodness-of-fit test, the Anderson-Darling test (AD), which checks how closely the points (here the residuals) follow a normal distribution. The high $p$ value ($p = 0.491$) indicates that there is no cause to reject the null hypothesis of normality. Thus, we safely assume normality.

Figure 10.4 illustrates MINITAB results from tests for homogeneity of error variances. Although the sample variances in table 10.9 showed some differences among groups, there is insufficient evidence to conclude that the population variances are different. Two different analyses, Bartlett's test and Levine's test, each give high $p$ values for the null hypothesis of equal error variances. Thus we are safe to assume homogeneity of variances for the stickleback data.

**Figure 10.3** MINITAB output for checking normality assumption. "RESI1" is the set of residuals generated from ANOVA on the stickleback data (example 10.2)

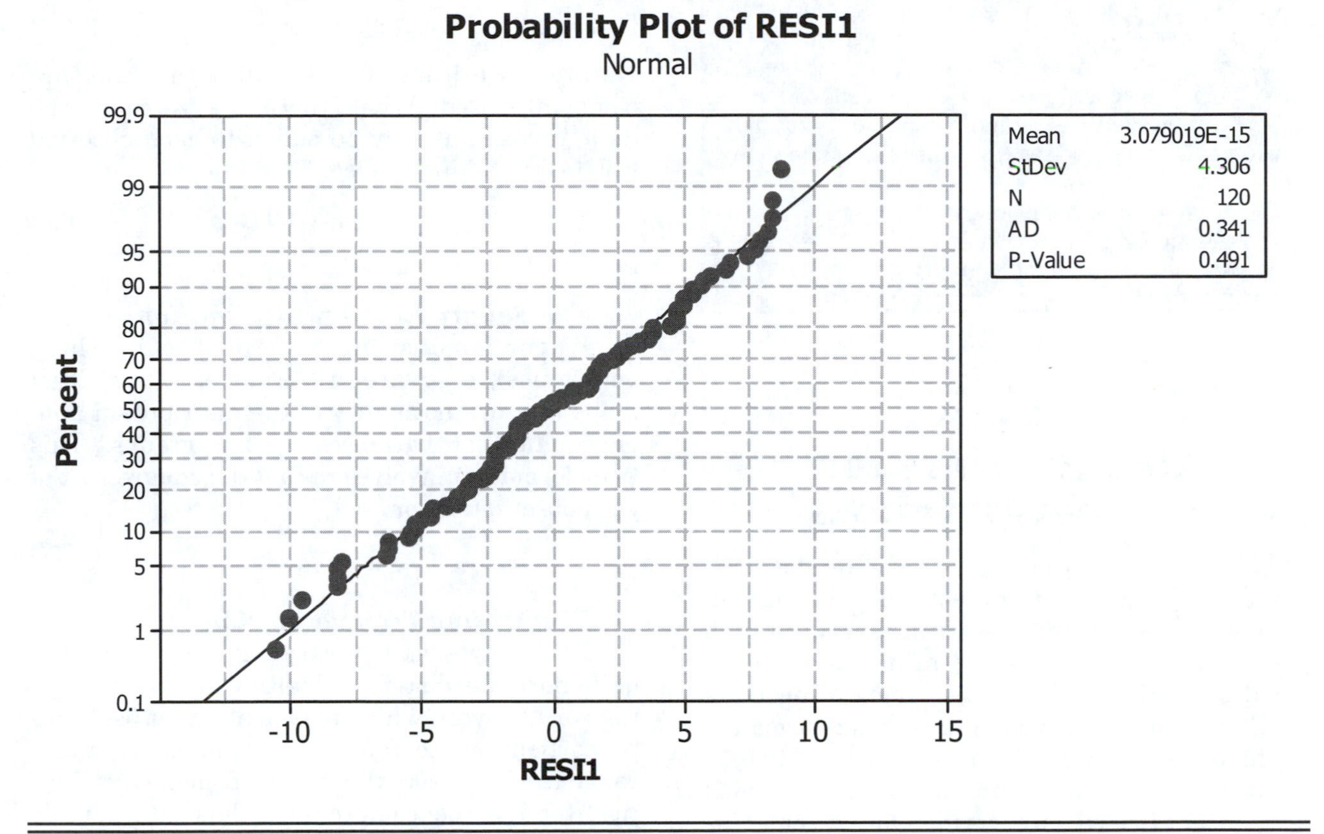


**Figure 10.4** MINITAB output for checking homogeneity of variances assumption for the stickleback data (example 10.2)

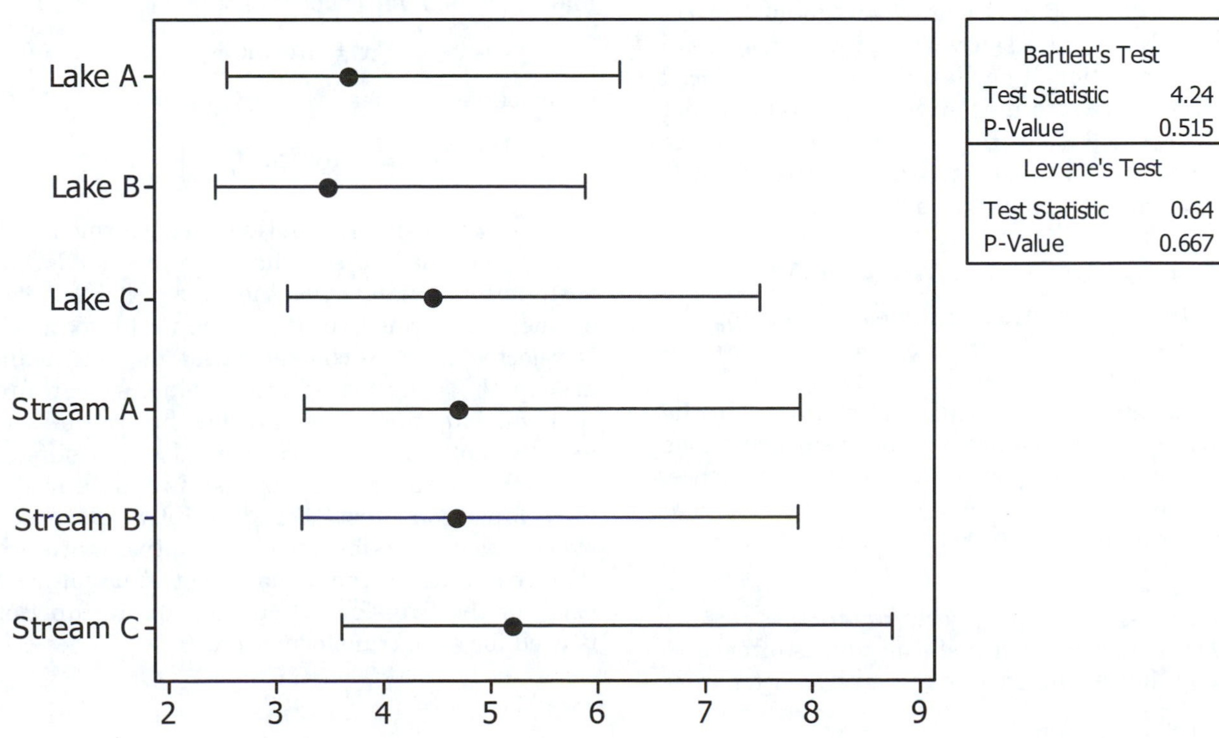

> **Caution**
>
> There are a number of important assumptions of ANOVA, one of which is that the within-group (error) variances of the various groups are all the same. ANOVA is said to be robust to the assumption of normality, meaning that some departure from this assumption is not too serious, particularly when large samples are involved. But ANOVA is very sensitive to the assumption of homogeneity of variances. When this assumption fails, ANOVA may give us the wrong answer and so should not be used.

## 10.5 Remedies for Failed Assumptions

There are situations in which the data collected in an experiment do not meet the normality and equal-variance assumptions, even though the variable in question is continuous and measured on an interval or ratio scale. When these problems exist, a mathematical transformation might correct the problem (see section 10.5.1). When either of these assumptions is not satisfied by the transformed data or when measurement is on an ordinal scale, one should use a nonparametric test. For the one-way ANOVA design, the nonparametric analogue is the Kruskal-Wallis Test (section 10.5.2). In no case should a parametric ANOVA be used with ordinal or nominal data! As mentioned earlier, the error introduced by using a discrete measurement variable is not large provided that the number of values the variable can assume is fairly large. This is assured when sample sizes are large.

### 10.5.1 Transformations in ANOVA

A **transformation** involves performing the same mathematical operation on each observation in a set of data. (This is easily done in a spreadsheet or with statistical software.) In the following sections, some common transformations and their uses are discussed. For convenience, original observations are designated by the usual $x$, and transformed observations are designated by $x'$.

***The Logarithmic Transformation***

The logarithmic transformation is often useful in ANOVA and in linear regression (chapter 12) where error variances are not equal. The log transformation is conducted by taking the logarithm of each observation, or

$$x' = \log(x) \tag{10.5}$$

Natural logarithms or base-10 logarithms are commonly used. When there are zeros in the data, a constant may be added to each observation before taking the log. For example,

$$x' = \log(x + 1) \tag{10.6}$$

since log(0) is undefined.

***The Square-Root Transformation***

When data occur in the form of counts (integer values), a square-root transformation often helps improve normality of the distribution and/or equality of variances among groups. This transformation involves taking the square root of each observation, or

$$x' = \sqrt{x} \tag{10.7}$$

***The Arcsine Transformation***

Percentages and proportions tend to be not quite normally distributed, although the underlying variable with which they deal might be. Since proportions range from 0 to 1 and percentages range from 0 to 100, the "tails" of such a distribution tend to be compressed. This situation is notable when many of the observations fall below 30% (0.3) or above 70% (0.7). This problem can often be rectified by an arcsine ($\text{sine}^{-1}$) transformation, which, for proportions, is

$$x' = \text{arcsine}\, x \tag{10.8}$$

or for percentages is

$$x' = \text{arcsine}\left(\frac{x}{100}\right) \tag{10.9}$$

Following transformation, transformed values are checked to see that they meet ANOVA assumptions. If all is OK, then the ANOVA is run on the transformed values. If the null hypothesis is rejected and we conclude that not all means are equal for the transformed value, we can further conclude that the means for the original values are not all equal. We can also use similar logic when interpreting the results from a multiple-comparisons test. For descriptive purposes, we can back transform the descriptive statistics. (For example, if the square-root function was used for the transformation, the square function is used for back transformation.)

### 10.5.2 Nonparametric Alternative to One-Way ANOVA: The Kruskal-Wallis Test

When data are measured on an ordinal scale, or when data are measured on an interval or ratio scale but the assumptions for ANOVA are not otherwise met, one may use a nonparametric test to compare three or more groups. These tests are often called nonparametric ANOVA tests, but since we do not measure variance in nonparametric tests, the name is a bit misleading. In effect, these tests tell us if it is reasonable to assume that three or more samples could have been drawn from identical populations.

The Kruskal-Wallis one-way ANOVA is the nonparametric counterpart of the one-way parametric design. It is used to test the null hypothesis that three or more independent samples were drawn from identical populations.

#### • Example 10.3

The Kruskal-Wallis test

The time (in seconds) that males of a certain species of grasshopper remained mounted on females during mating was determined. Ten males each were randomly assigned to one of four treatments (conditions). The data are as follows.

| Treatment A | Treatment B | Treatment C | Treatment D |
|---|---|---|---|
| 9 | 30 | 3,900 | 5 |
| 3 | 30 | 10,800 | 9 |
| 10 | 480 | 28,900 | 20 |
| 200 | 900 | 3,600 | 180 |
| 1 | 2 | 200 | 15 |
| 2 | 1 | 120 | 20 |
| 21 | 5,400 | 500 | 2 |
| 720 | 1,500 | 600 | 17 |
| 1,500 | 480 | 1,980 | 30 |
| 60 | 3 | 160 | 8 |

Modified data from R. Bland

The sample means, medians ($m$), and variances for the 4 treatments are:

| | Treatment A | Treatment B | Treatment C | Treatment D |
|---|---|---|---|---|
| $\bar{x}$ | 252.6 | 882.6 | 5076.0 | 30.6 |
| $m$ | 15.5 | 255 | 1290 | 16 |
| $s^2$ | 242,064 | 2,765,569 | 80,766,169 | 2,830 |

Although the experiment is properly designed for a completely randomized one-way ANOVA and the data are measured on a ratio scale, there are a couple of assumptions of the parametric test that these data do not satisfy. First, the variable does not seem to be even approximately normally distributed, as indicated by the large differences in the means and the medians of each group. As you will recall from chapter 6, the mean and median in a normal distribution are the same, and in samples drawn from such populations one would expect that these two values would be at least close to each other. Secondly, the within-groups variances are clearly not equal. (And a check of the homogeneity assumption with MINITAB confirms this suspicion.) Accordingly, we use the Kruskal-Wallis test.

**Assumptions of the Test**

1. The sampled populations have the same but unspecified distribution, with the possible exception that one or more of the sampled populations tends to have larger values than one or more of the others.
2. The samples represent random samples from their respective populations.
3. Measurement is on at least an ordinal scale.
4. The samples are from independent populations.

To conduct the Kruskal-Wallis test, all of the observations in all of the groups are ranked with respect to each other, with the lowest value in any of the samples receiving the rank of one. Tied values receive the average rank that they would have if they were not tied, as in previous examples. The ranked data and the sum of ranks for each group for the example are as follows.

| | Group A | Group B | Group C | Group D |
|---|---|---|---|---|
| | 10.5 | 19 | 37 | 8 |
| | 6.5 | 19 | 39 | 10.5 |
| | 12 | 27.5 | 40 | 15.5 |
| | 25.5 | 32 | 36 | 24 |
| | 1.5 | 4 | 25.5 | 13 |
| | 4 | 1.5 | 22 | 15.5 |
| | 17 | 38 | 29 | 4 |
| | 31 | 33.5 | 30 | 14 |
| | 33.5 | 27.5 | 35 | 19 |
| | 21 | 6.5 | 23 | 9 |
| $\Sigma R$ | 162.5 | 208.5 | 316.5 | 132.5 |

The test statistic used in the Kruskal-Wallis test is designated as $H$, which, when the null hypothesis that all of the samples were drawn from identical populations is true, has a chi-square distribution. A chi-square distribution is depicted in figure 10.5.

**Figure 10.5** A chi-square distribution

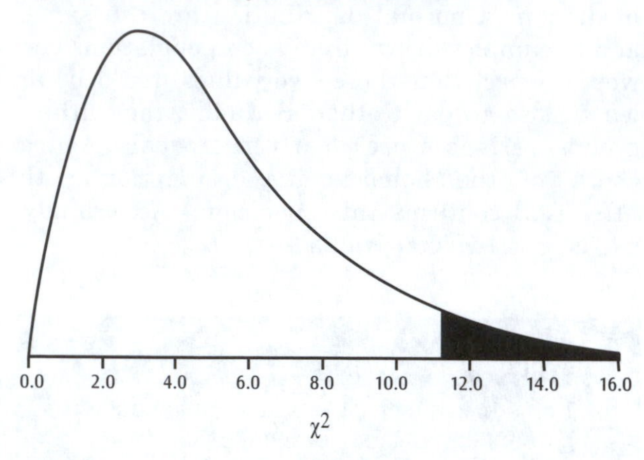


*H* is computed by

$$H = \frac{12}{n_t(n_t+1)}\left(\Sigma\frac{(\Sigma R_i)^2}{n_i}\right) - 3(n_t+1) \qquad (10.10)$$

where $n_i$ = number of observations in a group, $\Sigma R_i$ = the sum of ranks in a group, and $n_t$ = the total number of observations in all of the groups.

For the example

$$H = \frac{12}{40(40+1)}\left(\frac{(162.5)^2}{10} + \frac{(208.5)^2}{10} + \frac{(316.5)^2}{10} + \frac{(132.5)^2}{10}\right) - 3(40+1)$$

and after crunching the numbers:

$$H = 14.273$$

*H* has a chi-square distribution with degrees of freedom equal to the number of groups minus one. Consulting table A.3, we find that our calculated value of *H* is greater than the critical value of chi-square with 3 df at the 0.05 level (7.81). The *p* value is $.001 < p < .01$. We thus may reject $H_0$ and conclude that the four samples could not have been drawn from identical populations.

Doing a Kruskal-Wallis test with paper and calculator is quite tedious. Ranking is considerably easier using a spreadsheet (hint: put all the measurements in one column, group labels in a second column, sort by measurements, and put the numbers {1,2,3, . . . , $n_t$} in a third column). This is a good exercise once or twice, but then we can rely on statistical software to do these statistics on a routine basis. Results from MINITAB are displayed below.

**Table 10.10** MINITAB Kruskal-Wallis test results for the grasshopper mounting data

**Kruskal-Wallis Test:**
**TimeMounted versus Condition**

```
Kruskal-Wallis Test on TimeMounted
Condition     N    Median  Ave Rank      Z
Treatment A  10     15.50      16.3  -1.33
Treatment B  10    255.00      20.9   0.11
Treatment C  10   1290.00      31.7   3.48
Treatment D  10     16.00      13.3  -2.26
Overall      40                20.5

H = 14.27  DF = 3  P = 0.003
H = 14.29  DF = 3  P = 0.003 (adjusted for
ties)
```

MINITAB reports the same statistic as we calculated above and the exact *p* value shown falls within the range we determined from the chi-square table.

In this chapter we examined the one-way analysis of variance (ANOVA) and the multiple-comparisons test, which follows rejecting the null hypothesis of a fixed-effects model. We also explored how to check the assumptions ANOVA, and used two approaches for dealing with data that fail to meet these assumptions. In the following chapter, we examine two other widely used ANOVA layouts, the randomized block and factorial designs.

## Key Terms

among-groups variance (treatment variance)
analysis of variance (ANOVA)
fixed-effects (model I) ANOVA
multiple comparisons
random-effects (model II) ANOVA
transformation
Tukey test
within-group variance (error variance)

## Exercises

For each of the following problems, identify the ANOVA design involved, state the null hypothesis or hypotheses, and use the appropriate computations to construct an ANOVA table. Conduct a multiple-comparisons test, if appropriate to the specific problem. State a biological conclusion based on your statistical analysis. (Exercises preceded by an asterisk are suitable for a computer solution.)

Exercises 10.1 through 10.10 use the completely randomized one-way ANOVA design. You may assume a normal distribution and equality of within-group variances.

10.1 15 tobacco plants of the same age and genetic strain were randomly assigned to 3 groups of 5 plants each. One group was untreated, one was infected with tobacco mosaic virus (TMV), and one was infected with tobacco ringspot virus (TRSV). After one week the activity of odiphenol oxidase was determined in each plant. Does infection by either virus affect the activity of this enzyme?

Enzyme Activity ($\mu$l $O_2$/mg protein/min)

| Control | TMV-Infected | TRSV-Infected |
|---|---|---|
| 1.47 | 2.44 | 2.87 |
| 1.62 | 2.31 | 3.05 |
| 1.06 | 1.98 | 2.36 |
| 0.89 | 2.76 | 3.21 |
| 1.67 | 2.39 | 3.00 |

10.2 18 freshwater clams were randomly assigned to 3 groups of 6 each. One group was placed in the pond water from which the clams were collected, one group was placed in deionized water, and one group was placed in a solution of 0.5 mM sodium sulfate. At the end of a specified time period, blood potassium levels were determined. Do the treatments affect blood potassium levels ($\mu$ $K^+$)? (Data from J. Schiede).

| Pond Water | Deionized Water | Sodium Sulfate |
|---|---|---|
| .518 | .318 | .393 |
| .523 | .342 | .415 |
| .495 | .301 | .351 |
| .502 | .390 | .390 |
| .525 | .327 | .385 |
| .490 | .320 | .397 |

10.3 Cellulase activity (units/mg protein) in 4 genetic variants of a fungus species was measured in 5 randomly selected cultures of each variant. Is there a difference in activity of this enzyme in the four variants?

| Variant A | Variant B | Variant C | Variant D |
|---|---|---|---|
| 10 | 20 | 15 | 5 |
| 12 | 21 | 18 | 3 |
| 9 | 19 | 13 | 6 |
| 11 | 23 | 14 | 6 |
| 10 | 18 | 12 | 4 |

10.4 40 adult mice of a highly inbred laboratory strain were randomly assigned to 5 groups of 8 each in a controlled environment that was lit from 7 AM until 7 PM and dark from 7 PM until 7 AM. After several weeks plasma sodium concentration (meq/liter) was determined for one group at each of the times indicated in the data table. Does time of day influence plasma sodium concentration? (Data from J. Schiede.)

| 7 AM | 1 PM. | 7 PM | 1 AM | 7 AM |
|---|---|---|---|---|
| 148 | 138 | 140 | 137 | 143 |
| 152 | 130 | 138 | 135 | 152 |
| 150 | 132 | 129 | 148 | 149 |
| 149 | 129 | 140 | 140 | 150 |
| 160 | 140 | 142 | 137 | 161 |
| 155 | 139 | 139 | 141 | 155 |
| 162 | 130 | 132 | 145 | 149 |
| 149 | 142 | 141 | 136 | 162 |

10.5 15 juniper pythons of similar size and age were randomly assigned to 3 groups. One group was treated with drug A, one group with drug B, and the third group was not treated. Their systolic blood pressure was measured 24 hours after administration of the treatments. Does either drug affect blood pressure? Does one have more or less of an effect than the other?

| Drug A | Drug B | Untreated |
|---|---|---|
| 118 | 105 | 130 |
| 120 | 110 | 135 |
| 125 | 98 | 132 |
| 119 | 106 | 128 |
| 121 | 105 | 130 |

10.6 30 pea seeds were randomly assigned to 3 groups of 10 each. One group was germinated in the presence of chemical A, a second group with chemical B, and the third group served as a control. After five days, the length (in millimeters) of the primary root was measured. We wish to know if either or both of these chemicals affects root growth, and if so, if one is more effective than the other.

| Chemical A | Chemical B | Control |
|---|---|---|
| 115 | 120 | 82 |
| 103 | 125 | 97 |
| 98 | 122 | 105 |
| 121 | 100 | 90 |
| 130 | 90 | 102 |
| 107 | 128 | 98 |
| 106 | 121 | 105 |
| 120 | 115 | 89 |
| 100 | 130 | 100 |
| 125 | 120 | 90 |

10.7 Random samples of a certain species of zooplankton were collected from 5 randomly selected lakes and their selenium content was determined. We wish to know if there is a difference among lakes with respect to selenium content in this species (i.e., is there a significant "lake effect"?).

| Lake A | Lake B | Lake C | Lake D | Lake E |
|---|---|---|---|---|
| 23 | 34 | 15 | 18 | 25 |
| 30 | 42 | 18 | 15 | 20 |
| 28 | 39 | 12 | 9 | 22 |
| 32 | 40 | 10 | 12 | 18 |
| 35 | 38 | 8 | 10 | 30 |
| 27 | 41 | 16 | 17 | 22 |
| 30 | 40 | 20 | 10 | 20 |
| 32 | 39 | 19 | 12 | 19 |

*10.8 27 chickens were randomly assigned to 3 groups of 9 each. One group was fed chicken feed containing 1 ppb of a pesticide, a second group feed containing 100 ppb of the pesticide, and the third group feed containing none of the pesticide (control). Unfortunately, one of the animals in the 100 ppb group escaped and has not been seen since. After six weeks with these diets, SRBC antibody titers of the animals were determined. Is there an effect of the pesticide at either concentration? Note: because the data are unbalanced ($n$ not all equal), the formulas given in the text do not work—solve this one on the computer.

| 1 ppb | 100 ppb | Control |
|---|---|---|
| 5 | 9 | 7 |
| 8 | 10 | 11 |
| 5 | 11 | 12 |
| 12 | 9 | 9 |
| 11 | 13 | 8 |
| 6 | 10 | 11 |
| 7 | 11 | 8 |
| 5 | 10 | 9 |
| 6 | — | 10 |

10.9 An experiment similar to the one described in exercise 10.8 was conducted. In this case 4 randomly selected groups of 9 chickens each were given feed containing either one of three insecticides or none (control), and their *Brucella abortus* antibody titers were measured. We wish to know if exposure to insecticides affects the immune response to *B. abortus* (a bacterial pathogen) in chickens.

| Insecticide A | Insecticide B | Insecticide C | Control |
|---|---|---|---|
| 15 | 11 | 9 | 10 |
| 12 | 9 | 11 | 8 |
| 18 | 8 | 8 | 9 |
| 12 | 8 | 11 | 10 |
| 17 | 10 | 9 | 11 |
| 11 | 8 | 8 | 9 |
| 12 | 9 | 10 | 9 |
| 13 | 10 | 9 | 11 |
| 15 | 8 | 9 | 8 |

10.10 15 turkey hens were randomly assigned to 3 groups of 5. One group was given diet A, the second group diet B, and the third group diet C. We wish to know if there is a difference in the weight of eggs produced by birds on these diets, and if so, which diet results in the largest eggs. The data are the mean weights of 10 eggs from each bird.

| Diet A | Diet B | Diet C |
|---|---|---|
| 124 | 98 | 116 |
| 118 | 100 | 97 |
| 120 | 95 | 100 |
| 127 | 102 | 89 |
| 115 | 105 | 98 |

*10.11 Using the data from exercise 10.1, determine if the error variances may be considered to be equal.

*10.12 Using the data from exercise 10.2, determine if the error variances may be considered to be equal.

*10.13 Using the data from exercise 10.3, determine if the error variances may be considered to be equal.

*10.14 Using the data from exercise 10.4, determine if the error variances may be considered to be equal.

*10.15 Using the data from exercise 10.5, determine if the error variances may be considered to be equal.

10.16 Using the data from exercise 10.6, determine if the error variances may be considered to be equal.

10.17 Using the data from exercise 10.20, determine if the error groups variances may be considered to be equal.

10.18 Using the data from exercise 10.21, determine if the error groups variances may be considered to be equal.

Exercises 10.19 through 10.24 are appropriate for the Kruskal-Wallis test.

10.19 15 inbred laboratory rats were randomly assigned to 3 groups of 5 each. Each rat was inoculated with *Staphylococcus mucans*, a bacterium involved in the process of tooth decay. One group (control) was given no additional treatment. A second group was given drug A, and a third group was given drug B. At the end of six weeks, the extent of tooth decay in the rats was evaluated, as percent of tooth decay. This is a fairly subjective measurement based on the combined evaluation of three observers and should probably be regarded as an ordinal measurement. We wish to know if there is a difference among the three groups.

| Control | Drug A | Drug B |
|---|---|---|
| 87 | 63 | 45 |
| 76 | 70 | 60 |
| 65 | 87 | 43 |
| 81 | 92 | 56 |
| 75 | 70 | 60 |

10.20 30 chickens of a highly inbred strain were randomly assigned to 3 groups of 10. One group was injected with 100 $\mu$g norepinephrin per kilogram of body weight, a second group was injected with 100 $\mu$g epinephrin per kilogram of body weight, and a third group served as the control. SRBC plaque-forming cells per $10^6$ spleen cells were measured after a specified time interval. The variances among the three groups are not equal, and the variable may or may not be normally distributed. Is there a difference among the three groups?

| Norepinephrin | Epinephrin | Control |
|---|---|---|
| 15 | 70 | 535 |
| 155 | 45 | 370 |
| 110 | 95 | 420 |
| 90 | 95 | 315 |
| 35 | 70 | 485 |
| 100 | 315 | 230 |
| 30 | 140 | 370 |
| 40 | 260 | 320 |
| 75 | 230 | 335 |
| 105 | 400 | 475 |

10.21 Testosterone levels in mature roosters of three strains of chickens were measured. It appears that the variances within the groups are unequal, and we do not know if the variable is normally distributed. Is there a difference among strains?

| Strain A | Strain B | Strain C |
|---|---|---|
| 439 | 102 | 107 |
| 568 | 115 | 99 |
| 134 | 98 | 102 |
| 897 | 126 | 105 |
| 229 | 115 | 89 |
| 329 | 120 | 110 |

10.22 The aggressiveness of 15 randomly selected female gully cats was measured under 3 different conditions. Aggressiveness was measured on a rather subjective scale of 0 to 100 and should be regarded as an ordinal measurement. We wish to know if there is a difference in aggressiveness among the three conditions.

| Condition A | Condition B | Condition C |
|---|---|---|
| 5 | 20 | 60 |
| 10 | 15 | 50 |
| 20 | 20 | 70 |
| 8 | 25 | 25 |
| 11 | 27 | 90 |

10.23 3 groups of eggs of a tropical frog were exposed to either water containing no benzene (control group), water containing 50 ppm benzene, and water saturated with benzene. After hatching, the brain size of the froglets was measured. We wish to determine if exposure to benzene affects brain size. We do not know if this variable is approximately normally distributed, nor is it possible to determine its distribution with a sample as small as this. The results were as follows. (Data from J. Martin.)

| Control | 50 ppm | Saturated |
|---|---|---|
| 81 | 88 | 111 |
| 72 | 89 | 109 |
| 68 | 92 | 133 |
| 87 | 107 | |
| | 101 | |
| | 91 | |

10.24 24 rug rats were randomly assigned to 4 groups of 6 each. The groups were subjected to different levels of stress (high, moderate, low, and none) induced by occasional, unexpected loud noises. After one week, blood lymphocytes (cells $\times$ $10^6$/ml) were measured. We do not know if this variable is normally distributed in this species.

| High | Moderate | Low | None |
|---|---|---|---|
| 2.9 | 3.7 | 5.2 | 6.8 |
| 1.8 | 4.6 | 6.2 | 7.9 |
| 2.1 | 4.2 | 5.7 | 7.1 |
| 1.5 | 2.8 | 4.6 | 6.5 |
| 0.9 | 3.2 | 5.1 | 6.3 |
| 2.8 | 4.5 | 4.9 | 7.6 |

# 11

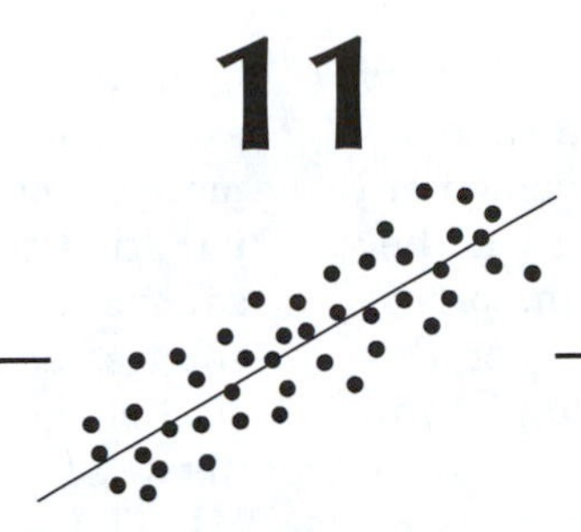

# Other ANOVA Designs

In chapter 10, we explored one-way analysis of variance (ANOVA) and its nonparametric analogue, the Kruskal-Wallis test. The one-way design looks at the effect of changing a single categorical factor (treatment) on a measurement of interest (response). The design is also completely randomized, in that subjects are randomly chosen to receive particular treatments. ANOVA can be extended to many other experimental designs. In this chapter, we briefly examine two of them: the randomized block design and the factorial design. Since the hand calculations are quite tedious, we'll use computer software to do all the number crunching. Our goals are to make sense of the statistical results and to be aware of certain design principles to guide us past any pitfalls.

## 11.1 The Randomized Block Design

Recall example 10.1, where we fed groups of mice one of three different diets and then measured their weight gain. We deliberately chose a strain of inbred mice to eliminate effects of genetic differences, and had a clear record of their history in the lab, so we could pick even-aged individuals for the experiment.

Suppose that we are interested in a wild population of white-footed deer mice rather than in highly inbred laboratory mice. We might expect that genetic variability would be fairly high in such a population. Furthermore, we would not know their ages. Using the completely randomized design (section 10.3), the additional variance in growth rates due to genetic and age differences among the experimental units would add to the error variance. This larger error variance would have the effect of reducing the variance ratio and making it more difficult to reject the null hypothesis (i.e., reduced power of the test).

The **randomized block design** provides a way to deal with this problem. In this design individuals are "blocked" (grouped) according to the characteristic whose variance we wish to identify and "partition out." In the case of our experiment with wild mice we would group individuals according to litters, since it is likely that members of the same litter are genetically similar and of the same age. Note that each individual observation (each mouse) is classified according to two criteria (diet and litter) and that only one individual occupies each possible combination of diet and litter. Thus, this particular design is sometimes called a two-way ANOVA without replication. It is a mixed-effects model, where diet is a fixed factor and litter is a random factor.

As in the completely randomized one-way ANOVA design, the null hypothesis in a randomized block design is that the treatment means are equal. We are also interested in knowing whether block (in this case, litter) is also an important source of variation. The null hypothesis for this second question is that block effect is unimportant.

## Example 11.1

### A Randomized Block Design

Ten litters of white-footed deer mice of approximately the same age were selected. One member of each litter was randomly assigned to one of the three treatment groups (the diets of example 10.1), and their weight gain was determined as before. The results are given in table 11.1.

**Table 11.1** Weight gain by white-footed deer mice fed mouse food (A), junk food (B), and health food (C)

| | | Treatments | | |
|---|---|---|---|---|
| | Litter Number | Group A | Group B | Group C |
| | 1 | 11.8 | 13.6 | 9.2 |
| | 2 | 12.0 | 14.4 | 9.6 |
| | 3 | 10.7 | 12.8 | 8.6 |
| | 4 | 9.1 | 13.0 | 8.5 |
| Blocks | 5 | 12.1 | 13.4 | 9.8 |
| | 6 | 9.8 | 10.9 | 10.0 |
| | 7 | 10.5 | 13.6 | 9.2 |
| | 8 | 10.5 | 11.9 | 8.8 |
| | 9 | 9.0 | 10.5 | 6.9 |
| | 10 | 11.2 | 13.8 | 10.1 |
| $\bar{x}$ | | 10.67 | 12.79 | 9.07 |

Several features about this design require attention. Treatments, in this case, are the different diets and represent the factor in which we have our primary interest. Treatments are customarily shown as separate columns. Blocks, which are the individual litters, represent the source of variation we regard as extraneous and wish to remove (or partition out) from what would otherwise be a part of the error variance. Blocks are shown as rows. Each treatment contains one and only one member of each block. In the example we randomly select one member of each litter for each treatment. The total number of observations ($n_t$) in such a design is therefore the number of treatments (diets) × the number of blocks (litters). So, in this example, the total numbers of observations is 3 × 10 = 30.

The calculations for the randomized block design are similar to those for the completely randomized design, except that they are a bit more extensive in order to partition the variance into more components. For this design, we wish to partition the total sum of squares into a sum of squares associated with the treatments, a sum of squares associated with the blocks, and an error sum of squares. We will skip any further calculations of sums of squares, leaving that to MINITAB (table 11.2). The degrees of freedom are determined by the number of different treatments and the number of different blocks. For the current example, we have three diets and 10 litters. Total degrees of freedom are $df_t = n_t - 1$. For the example, $df_t = 30 - 1 = 29$. Column (treatment) degrees of freedom are $df_c$ = columns − 1. For the example, $df_c = 3 - 1 = 2$. Row (block) degrees of freedom are $df_r$ = rows − 1. For the example, $df_r = 10 - 1 = 9$. Error degrees of freedom are $df_e = df_t - df_c - df_r$. For the example, $df_e = 29 - 2 - 9 = 18$. Table 11.2 shows the ANOVA table for this example, as generated by MINITAB.

**Table 11.2** MINITAB results for example 11.1, effects of diet and litter on weight gain in wild mice. This example uses a randomized block design, which MINITAB runs using the two-way ANOVA routine

**Two-way ANOVA: WeightGain versus Diet, Litter**

```
Source  DF       SS       MS      F      P
Diet     2   69.643  34.8213  73.25  0.000
Litter   9   25.934   2.8815   6.06  0.001
Error   18    8.557   0.4754
Total   29  104.134

S = 0.6895  R-Sq = 91.78%  R-Sq(adj) = 86.76%
```

The F statistic (variance ratio) for diet treatments (73.25) is found by dividing the treatment mean square by the error mean square. (Remember that "mean square" refers to the same thing as a variance.) The critical value of $F$ is found in table A.6 for $\alpha = 0.05$ at 2 and 18 degrees of freedom. Since 18 is not tabulated, we'll use the values for 2 and 20 df ($F_{crit} = 3.49$) and for 2 and 15 df ($F_{crit} = 3.68$). Since our calculated $F$ value is much larger than both those numbers, we reject $H_0$ and conclude that diet has a significant effect on weight gain in white-footed deer mice. It is also possible to test for significant differences among blocks. For this hypothesis, $F = 6.06$, which we compare to the critical value at 9 and 18 degrees of freedom. $F_{.05,9,20} = 2.39$ and $F_{.05,9,15} = 2.59$. Again, since the $F$ statistic (6.06) is much larger than these numbers, the second hypothesis is also rejected. Substantial variation was factored out with the different litters. This provides guidance for future experiments that using a block design is the best way to go when working with these animals.

A very common use of the randomized block design is when one desires to make three or more repeated measurements on the same individual. Thus, it may be used in much the same way that the paired $t$-test is used in a before-after study (section 9.6), except that more than two measurements are involved. (The paired $t$-test is, in fact, a special case of the randomized block design.) When used in this way, the randomized block design is sometimes called "**repeated-measures ANOVA.**"

## Example 11.2

### A randomized block design using repeated measures

The convict cichlid is one of several fish species that exhibits biparental care of eggs and fry. An experiment was conducted to determine if male convict cichlids might spend more time in direct offspring care with fry than with eggs. Accordingly, the time eight males spent in this activity (in seconds) during a 15-minute observation period was determined for five consecutive days. The eggs hatched after two days and for several days after that the fry were attached to the substrate by sticky "pads" on their heads. The results are given in table 11.3.

**Table 11.3** Brooding time (sec/15 min) by male convict cichlids

| Male Number | Day 1 | Day 2 | Day 3 | Day 4 | Day 5 |
|---|---|---|---|---|---|
| 1 | 11.9 | 2.2 | 57.9 | 259.5 | 200.4 |
| 2 | 42.7 | 60.7 | 71.2 | 163.3 | 228.1 |
| 3 | 15.8 | 14.8 | 311.3 | 283.9 | 436.3 |
| 4 | 191.2 | 148.8 | 437.8 | 319.2 | 462.6 |
| 5 | 3.5 | 187.3 | 281.4 | 410.4 | 373.7 |
| 6 | 23.7 | 0.0 | 98.6 | 185.7 | 106.8 |
| 7 | 0.0 | 0.0 | 102.4 | 400.7 | 386.9 |
| 8 | 33.5 | 107.5 | 193.5 | 317.8 | 337.3 |

Data from D. Dickens

In this case, days are treatments (shown as columns) and individual males are blocks (shown as rows). The randomized block design is appropriate in this case because we wish to repeatedly measure the same males, and we wish to partition out the individual variation among males. You will note that the variation among males is considerable in this example and might well be large enough to mask any treatment effect if it were a part of the error variance. The computer solution for this example is given below in table 11.4. The results lead us to conclude that we may reject the first null hypothesis. Treatment means are clearly quite different, as shown by the very low $p$ value ($p < 0.001$). Inspection of the means suggests that days 1 and 2 were not significantly different from each other but were significantly lower than days 3, 4, and 5. (This can be confirmed with a multiple-comparisons test.) We thus conclude that males spend less time brooding eggs than brooding fry. The second null hypothesis is also rejected, supporting our observation that different males vary a lot in their time devoted to offspring. (Fishes are not too different from humans in this regard!)

**Caution**

In a randomized block ANOVA, an important assumption is that interaction between the treatments and blocks is absent. Interaction means that treatments might affect individuals in one block differently than individuals in another block, or that the combined effect of the treatment and block factors somehow produce a result that is greater than the sum of their individual effects. Unfortunately, there is no way to check this assumption with the randomized block layout. If we have reason to suspect there are interactions, we best use a two-way factorial design instead (section 11.2).

**Table 11.4** MINITAB results for example 11.2, brooding time by age of young (treatment) and individual male (block). This example uses the same two-way ANOVA procedure to address a repeated-measures design

**Two-way ANOVA: BroodTime versus YoungAge, Male**

```
Source     DF        SS        MS       F       P
YoungAge    4    513610    128402   24.24   0.000
Male        7    215656     30808    5.82   0.000
Error      28    148325      5297
Total      39    877590

S = 72.78       R-Sq = 83.10%       R-Sq(adj) = 76.46%

                         Individual 95% CIs For Mean Based on
                         Pooled StDev
YoungAge        Mean     -+---------+---------+---------+--------
Day 1         40.288     (----*----)
Day 2         65.163       (-----*----)
Day 3        194.263                   (----*-----)
Day 4        292.563                              (----*-----)
Day 5        316.513                                (-----*----)
                         -+---------+---------+---------+--------
                          0       100       200       300
```

## 11.2 The Factorial Design

In many situations two or more factors (treatments) interact with each other to produce effects beyond the sum of the effects of the two acting alone. In other words, factors may interact either synergistically or antagonistically. For example, many medications should not be taken together. Either drug taken alone might produce its desired effect, but when taken together they interact in some harmful way.

When **interaction** among two or more factors is suspected, the **factorial design** is appropriate. Note that the factorial design is similar in layout to the randomized block design, except that both factors represent treatments in which we have an interest and each combination of treatments (cell) consists of replicated observations. Quite often both treatments are fixed effects. Since each individual is classified according to two criteria—the two main treatments—and since there are several individuals (replicates) within each combination of the two treatments, this design is sometimes called two-way ANOVA with replication.

### Example 11.3

A factorial design ANOVA

Consider another mouse diet experiment similar to the two examples used earlier to illustrate the completely randomized (example 10.1) and the randomized block (example 11.1) designs. In this experiment, however, we wish to determine the effect of diet, the effect of stress, and the interaction of these two factors, if any, on weight gain. We could conduct two separate experiments—one on diet and one on stress—but this would give us no information on their possible interaction.

For this experiment, we select 32 highly inbred mice of the same age and sex (to minimize the error variance) and assign them at random to four groups. One group receives the potato chip-Twinkie-cola diet (junk food diet) and listens to rap music eight hours each day (high stress). Another group has the same diet but listens to baroque music for eight hours each day (low stress). A third group receives regular mouse food (control) and listens to rap music (high stress). The fourth group receives the control diet and listens to baroque music. (The musical bias of the authors is evident here!) All four groups are housed under identical conditions. Thus, the two variables under study are arranged in all four possible combinations. Each combination of treatments has eight replicate mice ($n = 8$). This is an example of a $2 \times 2$ factorial design.

We have an interest in two main effects (or treatments), diet and stress, and whether a possible interaction effect exists. We therefore test three null hypotheses.

1) Effect of stress

$$H_0: {}_L = {}_H \quad vs \quad H_a: {}_L \neq {}_H$$

2) Effect of diet

$$H_0: {}_C = {}_J \quad vs \quad H_a: {}_C \neq {}_J$$

3) Interaction effect

$H_0$: No interaction *vs* $H_a$: Interaction

Data from this experiment are shown in table 11.5. The 32 measurements of weight gain (response variable) are arranged in four groups, representing the four combinations of treatments. We also show the cell means, which are illustrated later. We use MINITAB to generate the necessary statistics rather than going through the cumbersome calculations.

**Table 11.5** Effect of diet and stress level on weight gain in mice. Weight gain is measured as mg increased in one week

| Diet | Stress level Low | High |
|---|---|---|
| **Junk food** | 132 | 157 |
| | 128 | 143 |
| | 142 | 162 |
| | 131 | 150 |
| | 135 | 149 |
| | 120 | 140 |
| | 139 | 159 |
| | 133 | 158 |
| | mean **132.5** | mean **152.25** |
| **Control** | 120 | 130 |
| | 131 | 142 |
| | 122 | 131 |
| | 129 | 124 |
| | 120 | 124 |
| | 119 | 131 |
| | 134 | 143 |
| | 123 | 131 |
| | mean **124.75** | mean **132** |

The data for ANOVA may also be arranged in another way, as shown in table 11.6. This data layout is the structure commonly used by statistical software such as MINITAB. In this layout, all the measurements of the response variable are listed in one column and codes are placed in two other columns to indicate which treatment condition (level) is associated with each measurement. For the present example, the data matrix has 32 rows, for the 32 total mice in the experiment.

The degrees of freedom for each main effect is the number of groups minus one. For the example, $df_{diet}$ and $df_{stress}$ are both $2 - 1 = 1$. The degrees of freedom for interaction is degrees of freedom for one main effect times the degrees of freedom for the other main effect. For the example, this is $df_{interaction} = 1 \times 1 = 1$. The total degrees of freedom is $n_t - 1$, which, for the example, is $32 - 1 = 31$. The error degrees of freedom is the total degrees of freedom minus the degrees of freedom for each mean effect and interaction, which, for the example, is $31 - 1 - 1 - 1 = 28$. The

**Table 11.6** Effect of diet and stress level on weight gain in mice

| Weight Gain | Diet | Stress | Weight Gain | Diet | Stress |
|---|---|---|---|---|---|
| 132 | J | L | 157 | J | H |
| 128 | J | L | 143 | J | H |
| 142 | J | L | 162 | J | H |
| 131 | J | L | 150 | J | H |
| 135 | J | L | 149 | J | H |
| 120 | J | L | 140 | J | H |
| 139 | J | L | 159 | J | H |
| 133 | J | L | 158 | J | H |
| 120 | C | L | 130 | C | H |
| 131 | C | L | 142 | C | H |
| 122 | C | L | 131 | C | H |
| 129 | C | L | 124 | C | H |
| 120 | C | L | 124 | C | H |
| 119 | C | L | 131 | C | H |
| 134 | C | L | 143 | C | H |
| 123 | C | L | 131 | C | H |

Diets: J—junk food, C—control
Stress levels: L—low, H—high

sums of squares (generated by MINITAB), degrees of freedom, and variances (mean squares, MS) are arranged in an ANOVA table, as shown in table 11.7. Double-check that the numbers make sense and agree with those above.

**Table 11.7** Two-way factorial ANOVA using MINITAB for example 11.3

**ANOVA: WeightGain versus Diet, Stress**

```
Factor     Type     Levels    Values
Diet       fixed         2      C, J
Stress     fixed         2      H, L

Analysis of Variance for WeightGain

Source       DF      SS      MS      F      P
Diet          1  1568.0  1568.0  32.45  0.000
Stress        1  1458.0  1458.0  30.17  0.000
Diet*Stress   1   312.5   312.5   6.47  0.017
Error        28  1353.0    48.3
Total        31  4691.5

S = 6.95136 R-Sq = 71.16% R-Sq(adj) = 68.07%
```

Raw data are shown in tables 11.5 and 11.6

As before, each MS is determined by dividing the SS by the associated df. The $F$ statistics are determined by dividing each MS by the $MS_{error}$. For example:

$$F_{stress} = \frac{MS_{stress}}{MS_{error}} = \frac{1458}{48.32} = 30.18$$

We compare the calculated $F$ ratios with the critical $F$ values from table A.6. Since 28 df is not tabulated, we find the critical values from the table that bracket 1 and 28 df. These would be $F_{1,25} = 4.24$ and $F_{1,30} = 4.17$.

Since all three $F$ statistics have the same number of degrees of freedom associated with them (table 11.7), the critical value is the same. Inspection of the table indicates that all three $F$ statistics exceed the critical value of 4.24, so we have grounds to reject all three null hypotheses. Using table A.6, we can find critical values only at $\alpha = 0.05$ and $\alpha = 0.01$, so using this table alone would only give us $p$ values of $p < 0.01$. However, notice that the $F$ statistics for the two main effects are much larger than the critical value at $\alpha = 0.01$, suggesting that their $p$ values are much less than 0.01. Indeed, MINITAB reports $p = 0.000$ (actually a very small number—not zero! $p < 0.001$) for each main effect (diet, stress). We thus can conclude that there are significant effects of both diet and stress on weight gain in these mice.

Finally, we may also reject the last $H_0$ (no interaction) and conclude that there is a significant interaction of diet and stress on weight gain. The highest weight gain was induced in individuals exposed to the junk food diet and high stress, and the lowest weight gain was induced in individuals exposed to the regular diet and low stress. Results of this nature are sometimes easier to interpret graphically, as shown in figure 11.1. This figure is an example of an **interaction plot**, which illustrates the means for each treatment combination. Connecting the means from different levels of factor 1 (in this case, stress) allows us to see if the response (change in mean with change in level) is similar for different levels of factor 2 (diet). No interaction is indicated when the lines drawn between the means are nearly parallel. In this case, the lines are not parallel so the weight gain due to junk food appears to be enhanced in a high-stress environment.

If you have computer software available, try running the ANOVA on this example and check that your ANOVA table shows similar $F$ statistics. Also, produce an interaction plot. Finally, check the assumptions of ANOVA (section 10.4). You should see that both the normality and homogeneous variance assumptions were met.

The factorial design is not restricted to two levels of each treatment, as in the preceding example, but may include as many levels of the treatments as desired by the investigator. Example 11.4 illustrates a factorial design using two levels of one main effect and three levels of a second main effect. This is called a 2 × 3 factorial design.

**Figure 11.1** An interaction plot. The graph illustrates the cell means for example 11.3

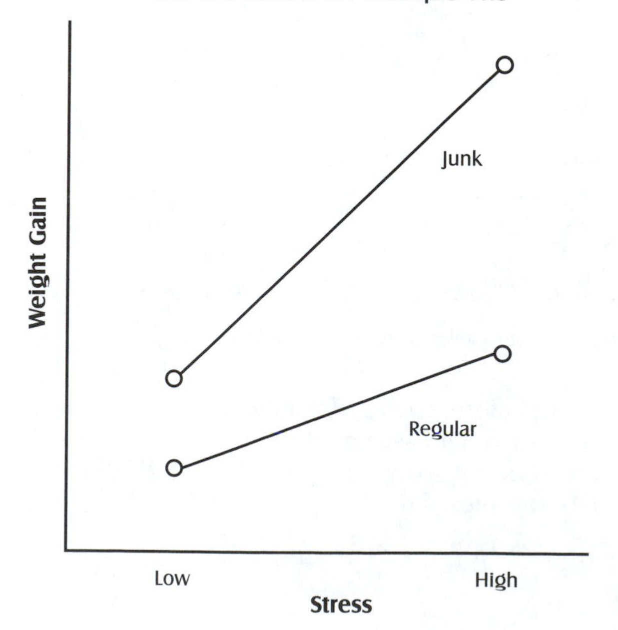


## Example 11.4

### A factorial design with more than two levels of a main effect

An evolutionary biologist selected samples of road warblers (museum specimens) of both sexes from three locations: Eastern North America, Western North America, and the Intermountain region of North America. Ten individuals, who were classified by these two criteria (sex and location) were selected at random, and their culmen (bill) lengths were measured. The results were as follows.

| | Eastern | Western | Intermountain |
|---|---|---|---|
| **Females** | 50.1 | 53.4 | 54.0 |
| | 52.8 | 55.2 | 49.1 |
| | 50.8 | 51.0 | 60.5 |
| | 58.8 | 59.3 | 57.8 |
| | 59.7 | 61.5 | 48.7 |
| | 49.0 | 61.2 | 57.0 |
| | 58.8 | 57.8 | 61.1 |
| | 62.2 | 50.1 | 62.8 |
| | 57.8 | 56.0 | 59.8 |
| | 61.2 | 56.5 | 60.3 |
| **Males** | 46.5 | 57.5 | 49.1 |
| | 44.4 | 59.3 | 51.8 |
| | 42.0 | 62.4 | 55.3 |
| | 51.1 | 61.1 | 43.6 |
| | 45.8 | 59.9 | 50.1 |
| | 46.3 | 55.6 | 51.0 |
| | 41.8 | 56.8 | 49.0 |
| | 52.0 | 59.2 | 48.8 |
| | 46.5 | 50.4 | 52.0 |
| | 39.0 | 47.8 | 43.0 |

**Table 11.8** Two-way factorial ANOVA for the bill-length data (example 11.4)

**ANOVA: BillLength versus Location, Gender**

```
Factor          Type     Levels      Values
Location       fixed          3     E, I, W
Gender         fixed          2        F, M

Analysis of Variance for BillLength

Source                DF          SS         MS        F        P
Location               2      335.94     167.97     8.67    0.001
Gender                 1      511.58     511.58    26.41    0.000
Location*Gender        2      350.84     175.42     9.06    0.000
Error                 54     1046.03      19.37
Total                 59     2244.39

S = 4.40124      R-Sq = 53.39%       R-Sq(adj) = 49.08%
```

The MINITAB solution to this problem is given in table 11.8. You should be able to interpret the results. Also, draw an interaction plot (like figure 11.1), plotting location on the $x$-axis and different lines for the two genders and interpret it.

## 11.3 The Friedman Test

The Friedman two-way ANOVA tests whether three or more *related* samples could have been drawn from identical populations. Thus, the Friedman two-way ANOVA is analogous to the randomized block design. The Friedman test is a nonparametric procedure.

To conduct the test, the data are arranged in a two-way table of $k$ columns, representing the treatments, and $n$ rows, representing the blocks. We will examine a data set, use MINITAB to do the hard part, and interpret these results. Essentially, MINITAB will do an analysis with ranks and sum them in different ways to generate a statistic that can be compared to a chi-square distribution. The logic of this test is that if there were no difference among the treatments (columns), the sums of the ranks of the columns would be approximately equal. This is because any rank would be as likely to occur in any column as in any other.

### Example 11.5

The Friedman two-way ANOVA

Do neonatal garter snakes exhibit a decremental response (habituation) to a repeated overhead stimulus? Six snakes were chosen at random and placed in the experimental chamber and allowed to acclimate for approximately 30 minutes. A rapidly moving object was then presented overhead at 10-second intervals, and the reaction of the snake was noted. Reactions were scored from 3 (a rapid retreat from the stimulus) through 0, which indicated no response.

These are ordinal data, so a nonparametric test must be used. Since each animal was tested repeatedly, the observations are related and not independent. The data of the experiment are shown in table 11.9.

**Table 11.9** Reaction scores of snakes (example 11.5)

| Snake Number (Block) | Interval (Treatment) 1 | 2 | 3 | 4 | 5 | 6 | 7 | 8 |
|---|---|---|---|---|---|---|---|---|
| 1 | 3 | 3 | 2 | 2 | 0 | 1 | 0 | 0 |
| 2 | 3 | 2 | 0 | 2 | 0 | 1 | 0 | 0 |
| 3 | 2 | 0 | 0 | 0 | 3 | 0 | 0 | 0 |
| 4 | 2 | 2 | 0 | 0 | 0 | 0 | 0 | 0 |
| 5 | 3 | 0 | 2 | 0 | 2 | 0 | 0 | 0 |
| 6 | 2 | 2 | 0 | 1 | 1 | 1 | 1 | 0 |

Data from R. Hampton and J. Gillingham

When $n$ and $k$ are not too small, the calculated test statistic has a chi-square distribution. We may therefore consult table A.3 to determine if the differences among the eight treatments are significant. The degrees of freedom in the Friedman test are $k - 1$, which, for the example, is $8 - 1 = 7$. The critical value of chi-square at alpha = 0.05 with 7 degrees of freedom is 14.07. Note that this test does not allow us to determine if the block effect is important.

**Table 11.10** Results of Friedman's test on reaction scores of snakes (example 11.5)

**Friedman Test: ReactionScore versus Interval blocked by Snake**

```
S = 17.44 DF = 7 P = 0.015
S = 22.96 DF = 7 P = 0.002 (adjusted for ties)
```

| Interval | N | Est Median | Sum of Ranks |
|---|---|---|---|
| I-1 | 6 | 2.328 | 45.5 |
| I-2 | 6 | 1.703 | 35.5 |
| I-3 | 6 | 0.328 | 23.0 |
| I-4 | 6 | 0.578 | 26.5 |
| I-5 | 6 | 0.453 | 27.0 |
| I-6 | 6 | 0.453 | 23.5 |
| I-7 | 6 | 0.266 | 19.0 |
| I-8 | 6 | 0.016 | 16.0 |

Grand median = 0.766

Since the calculated value of the test statistic ($S$ = 17.44) is greater than 14.07, we may reject the null hypothesis that there is no significant difference between the treatments. The low $p$ value (0.015) is evidence that the treatment effect is discernable. Since, in this case, "treatments" are the successive presentations of the test stimulus at 10-second intervals, we further conclude that neonatal garter snakes exhibit a decremental response (i.e., they habituate) to the stimulus.

## 11.4 Other ANOVA Designs

Analysis of variance (ANOVA) is highly versatile. Since its invention by R. A. Fisher, ANOVA has been widely applied in agricultural experiments and research in ecology. We have just barely scratched the surface on uses of this technique.

In chapter 10, we explore the layout and applications of one-way ANOVA and a related nonparametric procedure. In the current chapter, we explore the modification of ANOVA for related samples (the randomized block design) and the extension for analyzing the simultaneous effects of two factors on the response variable (two-way factorial ANOVA). The advantages of the two-way design are efficiency (run two experiments at once) and also the ability to check for interaction effects.

ANOVA can be applied to many other research designs. Factorial ANOVA designs have been extended to three and even four factors manipulated simultaneously (e.g., Box et al. 1978). Although informative, these experiments can be quite complicated to run and, even worse, difficult to interpret because of the many possible interaction effects. Interactions may cloud our ability to detect and interpret main effects. Additional experimental designs using ANOVA include the split plot, the Latin square, and nested ANOVA. ANOVA can also be combined with regression analysis to compare two or more regression lines, a procedure called analysis of covariance (discussed in section 12.11).

Students who expect to use ANOVA in their research are encouraged to read further in one of the advanced statistical texts, such as those listed in the references (appendix C).

### Key Terms

factorial design
interaction effect
interaction plot
randomized block design
repeated-measures ANOVA

### Exercises

For the following problems, unless instructed otherwise, assume that the residuals are normally distributed and variances homogeneous. For each problem, identify the ANOVA design involved, state the null hypothesis or hypotheses, and, using statistical software, construct an ANOVA table. State a biological conclusion based on your statistical analysis. Where appropriate, construct an interaction plot and interpret it. All exercises are suitable for a computer solution.

Exercises 11.1 through 11.8 use a randomized block design.

11.1 3 mice from each of 6 litters were randomly assigned to 3 treatment groups. One group was exposed to 100 ppb methyl mercury, one group was exposed to 100 ppb mercuric chloride, and the third group served as a control. The time (in minutes) that each mouse spent on an exercise wheel in one day was measured several days after exposure. Does exposure to mercury affect the mice activity levels? Do mice show variation among litters in activity levels?

| Litter Number | Methyl Mercury | Mercuric Chloride | Control |
|---|---|---|---|
| 1 | 60 | 90 | 50 |
| 2 | 100 | 120 | 80 |
| 3 | 40 | 45 | 35 |
| 4 | 120 | 110 | 100 |
| 5 | 80 | 105 | 60 |
| 6 | 130 | 155 | 105 |

11.2 An investigator wished to test the effect of 3 diets on the peristaltic blood pressure of mangrove toads (measured in mm Hg). Because these animals are rare, only a few may be collected in any one location. 3 toads were collected from each of 5 locations and one toad from each location was randomly assigned to one of the 3 diets. Is there a difference in blood pressure that is related to the 3 diets?

| Location | Diet A | Diet B | Diet C |
|---|---|---|---|
| 1 | 65 | 75 | 80 |
| 2 | 60 | 69 | 79 |
| 3 | 55 | 50 | 70 |
| 4 | 53 | 54 | 80 |
| 5 | 64 | 69 | 85 |

11.3 Photosynthesis in 5 tobacco plants was measured before, one day after, and one week after exposure at a temperature of 100°F for two hours. We wish to know if exposure to this temperature affects the rate of photosynthesis, and if so, if the rate of photosynthesis returns to its pretreatment level within one week.

| | Photosynthesis ($\mu$M $CO_2$/min/g) | | |
|---|---|---|---|
| Plant Number | Before Treatment | One Day After Treatment | One Week After Treatment |
| 1 | 127 | 107 | 130 |
| 2 | 130 | 111 | 127 |
| 3 | 240 | 222 | 250 |
| 4 | 116 | 98 | 120 |
| 5 | 215 | 201 | 200 |

11.4 Males of the northern water snake apparently court females at the expense of time spent foraging. Thus, the energetic "cost" of courtship, if any, might be measured as weight loss in males through the short mating season. Five marked male northern water snakes were captured, weighed, and released at the beginning, near the middle, and at the end of the mating season. Using the data given in the following table, determine if there was a significant weight loss in males during this period. (Experiment suggested by H. Carbone.)

| | Weight (in grams) | | |
|---|---|---|---|
| Snake Number | Beginning | Middle | End |
| 1 | 397 | 362 | 325 |
| 2 | 410 | 385 | 350 |
| 3 | 362 | 325 | 300 |
| 4 | 291 | 270 | 253 |
| 5 | 325 | 289 | 280 |

11.5 3 hens from each of 3 varieties were randomly assigned to 3 diets. We wish to determine if there is a difference in egg weight produced by the 3 diets, and if so, which diet produces the largest eggs. The data are the mean weights (in g) of 10 eggs from each bird.

| | Mean Egg Weight with | | |
|---|---|---|---|
| Variety | Diet A | Diet B | Diet C |
| 1 | 112 | 87 | 100 |
| 2 | 100 | 82 | 98 |
| 3 | 90 | 75 | 86 |

11.6 A parcel of land was divided into 6 equally sized plots (blocks), and each block was divided into 3 equally sized subplots. 3 treatments (no added nitrogen, 10 lbs nitrogen/hectare, and 100 lbs nitrogen/hectare) were randomly assigned to the subplots within each plot. Does added nitrogen increase the growth of a certain crop? Data are pounds of yield per subplot.

| | Crop Yield with | | |
|---|---|---|---|
| Block | No added N | 10 lbs/ hectare | 100 lbs/ hectare |
| 1 | 105 | 156 | 187 |
| 2 | 98 | 145 | 167 |
| 3 | 125 | 170 | 201 |
| 4 | 100 | 150 | 180 |
| 5 | 130 | 185 | 210 |
| 6 | 80 | 135 | 162 |

11.7 The concentration of unicellular algae (measured as chlorophyll in μg/L) at 3 different depths in 4 lakes was measured. We wish to know if there is a difference in algae concentration that is related to depth. Lakes are treated as blocks to take into account differences among lakes.

| Lake | Surface | 1 m | 3 m |
|---|---|---|---|
| 1 | 425 | 130 | 56 |
| 2 | 500 | 215 | 115 |
| 3 | 100 | 30 | 10 |
| 4 | 325 | 100 | 28 |

11.8 A plant ecologist wished to determine if an exotic weed might be becoming more numerous in a particular area. She randomly selected 5 1-$m^2$ quadrats in the area of interest and counted the number of individual plants of the weed in each quadrat over a period of 5 years. Is there a significant change in density of this species over this time period? By inspecting the data, does there appear to be an increase over time?

| | Year | | | | |
|---|---|---|---|---|---|
| Quadrat Number | 1 | 2 | 3 | 4 | 5 |
| 1 | 2 | 5 | 8 | 10 | 20 |
| 2 | 5 | 9 | 17 | 30 | 40 |
| 3 | 15 | 30 | 31 | 60 | 72 |
| 4 | 0 | 2 | 9 | 15 | 24 |
| 5 | 9 | 11 | 23 | 17 | 45 |

Exercises 11.9 through 11.12 use a two-way factorial design.

11.9 The possible influence of crowding and sex on plasma corticosterone (in ng/mL) in a highly inbred strain of rug rats was investigated using a factorial design. Sex (males, nongravid females, and gravid females) and crowding (low, moderate, and high) were used as the main treatment effects. Write a paragraph or two discussing the results from a biological perspective, and support your conclusions with the proper statistical analysis. Include a graphic representation of the effects of these two factors and any possible interaction between them.

| | Crowding | | |
|---|---|---|---|
| Sex | Low | Moderate | High |
| Males | 5 | 115 | 253 |
| | 8 | 122 | 249 |
| | 13 | 119 | 260 |
| | 9 | 130 | 257 |
| | 15 | 114 | 280 |
| | 11 | 129 | 263 |
| Nongravid Females | 12 | 112 | 219 |
| | 19 | 115 | 222 |
| | 15 | 121 | 218 |
| | 20 | 117 | 220 |
| | 11 | 118 | 223 |
| | 18 | 120 | 225 |
| Gravid Females | 37 | 157 | 289 |
| | 42 | 160 | 273 |
| | 50 | 173 | 280 |
| | 35 | 182 | 291 |
| | 40 | 168 | 205 |
| | 36 | 170 | 296 |

11.10 45 large female guppies were randomly assigned to 9 groups, each of which received different amounts of food and were kept at different temperatures in a two-way factorial design. The number of offspring that each produced in a single brood are given in the following table. We wish to know if there is an effect of temperature, an effect of food intake, and an interaction of these two factors on the number of offspring per brood produced by these animals.

| | Number of Daily Feedings | | |
|---|---|---|---|
| Temperature | 1 | 2 | 3 |
| 70°F | 18 | 25 | 28 |
| | 20 | 30 | 36 |
| | 15 | 19 | 29 |
| | 27 | 30 | 30 |
| | 30 | 25 | 37 |
| 75°F | 20 | 28 | 33 |
| | 28 | 29 | 39 |
| | 30 | 32 | 42 |
| | 17 | 38 | 47 |
| | 29 | 29 | 38 |
| 80°F | 35 | 35 | 51 |
| | 30 | 39 | 42 |
| | 32 | 30 | 48 |
| | 28 | 40 | 39 |
| | 35 | 38 | 55 |

11.11 The possible effects of sex and age on systolic blood pressure in hamsters were investigated, with the following results. Determine if there is an effect of sex, an effect of age, and an interaction of these two factors on blood pressure (in mm Hg).

| | Sex | |
|---|---|---|
| Age | Male | Female |
| Adolescent | 108 | 110 |
| | 110 | 105 |
| | 90 | 100 |
| | 80 | 90 |
| | 100 | 102 |
| Mature | 120 | 110 |
| | 125 | 105 |
| | 130 | 115 |
| | 120 | 100 |
| | 130 | 120 |
| Old | 145 | 130 |
| | 150 | 125 |
| | 130 | 135 |
| | 155 | 130 |
| | 140 | 120 |

11.12 Using the data from table B.3 (see digital appendix), determine if there is an effect of smoking, an effect of sex, and an interaction between these two factors on pulse rate in humans.

Exercises 11.13 through 11.14 are appropriate for the Friedman two-way ANOVA.

11.13 6 male mosquito fish were selected at random from a large population. Each was placed individually in an aquarium with a female, and the number of copulatory attempts made during 5 successive three-minute intervals was recorded. The data do not have homogeneous variances. We wish to know if males exhibit a diminishing response (habituate) to individual females.

| | Interval Number | | | | |
|---|---|---|---|---|---|
| Male Number | 1 | 2 | 3 | 4 | 5 |
| 1 | 20 | 10 | 7 | 5 | 2 |
| 2 | 18 | 7 | 5 | 5 | 3 |
| 3 | 25 | 13 | 9 | 4 | 1 |
| 4 | 10 | 5 | 3 | 2 | 1 |
| 5 | 12 | 15 | 8 | 3 | 0 |
| 6 | 17 | 4 | 5 | 2 | 1 |

11.14 3 randomly selected mice from each of 6 litters were randomly assigned to 3 treatment groups. One group received injections of 100 μg/kg of epinephrin, the second group received injections of 100 μg/kg of norepinephrin, and the third group served as a control. SRBC antibody titers were measured one week after treatment. The variable does not appear to be normally distributed, nor are the error variances equal.

| Litter Number | Epinephrin | Norepinephrin | Control |
|---|---|---|---|
| 1 | 125 | 180 | 350 |
| 2 | 225 | 112 | 400 |
| 3 | 180 | 290 | 375 |
| 4 | 300 | 225 | 495 |
| 5 | 115 | 325 | 427 |
| 6 | 98 | 115 | 510 |

Choose the appropriate ANOVA design for exercises 11.15 through 11.19

11.15 6 rug rats were given a small amount of caffeine. Their pulse rate (in bpm) was measured before, immediately after, and one hour after administration of the caffeine. Does caffeine affect pulse rate in this species?

| Rat Number | Before | Immediately After | One Hour After |
|---|---|---|---|
| 1 | 105 | 115 | 108 |
| 2 | 98 | 110 | 100 |
| 3 | 110 | 125 | 115 |
| 4 | 100 | 112 | 105 |
| 5 | 114 | 130 | 120 |
| 6 | 90 | 100 | 95 |

11.16 6 damselfly naiads were placed individually into 6 containers, which contained 10 each of 4 different prey species. After a short time, the number of each prey species eaten was recorded. Is there a preference for any of the prey items? Note that we expect these data not to have homogeneous variances.

| Naiad Number | Prey A | Prey B | Prey C | Prey D |
|---|---|---|---|---|
| 1 | 8 | 3 | 1 | 2 |
| 2 | 9 | 5 | 4 | 1 |
| 3 | 7 | 1 | 0 | 0 |
| 4 | 10 | 7 | 8 | 1 |
| 5 | 8 | 9 | 1 | 2 |
| 6 | 9 | 2 | 1 | 3 |

11.17 The effect of infection by tobacco mosaic virus (TMV) and tobacco ringspot virus (TRSV) on *o*-diphenol oxidase activity in 3 genetic strains of tobacco was measured. Answer the following questions and support your answers with the appropriate ANOVA terms:

1. Does virus infection affect this enzyme?
2. Is there a difference in the activity of this enzyme among strains?
3. Do different strains respond differently to virus infection (i.e., is there a significant interaction between the two main effects)?

Show the interaction results graphically.

| | Treatment | | |
|---|---|---|---|
| | Noninfected | TMV-infected | TRSV-infected |
| Strain A | 102 | 237 | 117 |
| | 115 | 219 | 95 |
| | 98 | 201 | 128 |
| | 97 | 193 | 105 |
| Strain B | 85 | 175 | 60 |
| | 63 | 160 | 91 |
| | 127 | 230 | 135 |
| Strain C | 150 | 249 | 145 |
| | 168 | 250 | 170 |

11.18 Using the data from table B.3, determine if reaction time is different between males and females, athletes and nonathletes; also determine if there is an interaction between sex and participation in a varsity sport.

11.19 Testosterone levels in 6 captive male bush hogs was measured at 4 times during the year. The mating season for this species occurs in the fall. Do the data below support the hypothesis that testosterone levels in this species are higher during the mating season?

| | Plasma Testosterone Level (ng/mL) | | | |
|---|---|---|---|---|
| Male Number | Winter | Spring | Summer | Fall |
| 1 | 20 | 30 | 25 | 220 |
| 2 | 30 | 20 | 70 | 210 |
| 3 | 40 | 90 | 50 | 230 |
| 4 | 35 | 40 | 40 | 190 |
| 5 | 60 | 60 | 105 | 100 |
| 6 | 55 | 40 | 72 | 210 |

# 12

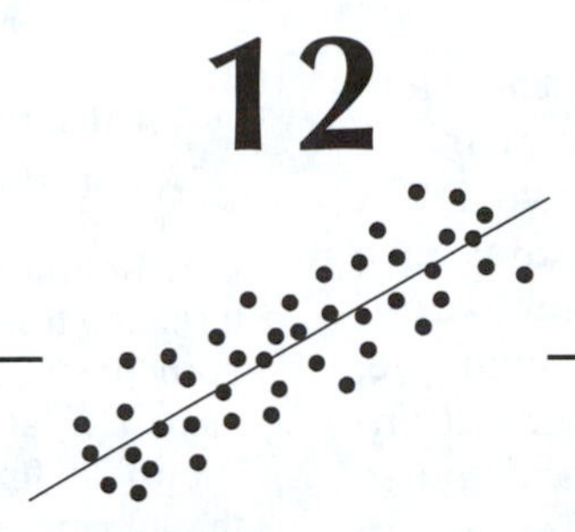

# Modeling One Measurement Variable against Another

## Regression Analysis

Often in biological research we are interested in exploring the possible relationships between two or more measurement variables. Examples include the relationship between body weight and blood pressure, the size of females and the number of offspring they produce, and the relationship between the dosage level of a drug and a particular physiological response. In all of these cases, two variables are measured for each individual in our sample, and we try to detect a relationship between these two variables. Notice the distinction from one-way ANOVA, where one of the variables (the treatment factor) is categorical and the other (the response) is a measurement variable.

In this chapter we consider statistical techniques designed to model the dependence of one variable on another (regression analysis). In chapter 13, we examine techniques for determining the strength of association between two variables (correlation). Which of these techniques is used for any particular problem depends on the nature of the data collected and on the nature of the questions being asked about the data.

### 12.1 Regression versus Correlation

**Correlation analysis** is used to determine (1) if an association between two variables exists and (2) how strong such an association is. By association, we mean that as one variable changes, the other changes in some consistent way. Note that in the case of correlation, there is no assumption about a cause-and-effect association between the two variables, although such a relationship might exist. For example, we might wish to know if students' scores on exams are associated with their scores on their take-home problem sets. In other words, do students who do well on problem sets also tend to do well on exams? We are not attempting to attribute causation of one variable from another. Both may instead be responding to other variables, such as class attendance or time spent working practice exercises. We only want to know how strongly they are associated.

**Regression analysis**, on the other hand, assumes a cause-and-effect relationship between the two variables such that at least a substantial proportion of the variation in one of the variables, called the **response variable**, can be explained by or attributed to the other variable, called the **predictor variable**. (These may also be called the **dependent variable** and **independent variable**, respectively.) For example, we might wish to investigate a possible relationship between caffeine consumption and pulse rate. We can assume that, if such a relationship exists, it is the amount of caffeine consumed (the independent variable) that causes a change in pulse rate (the dependent variable), and not the other way around!

Another important distinction between correlation and regression is that, in the most common type of regression analysis, the independent variable usually is not a normally distributed random variable. Rather, it is under the control of the investigator. In this respect regression as it is usually done is a fixed-effects design (model I), analogous to a fixed-effects ANOVA (see chapter 10). In effect this means that the experimenter chooses what levels of caffeine to administer to the test subjects in our example. A random-effects design (model II), analogous to a random-effects ANOVA, may also be used with regression analysis. However, this model is less common, so we do not detail it here.

Regression analysis may be extended to include more than one independent variable acting on a dependent variable, a technique known as multiple linear regression. However, our treatment here will be limited to simple linear regression, in which only one independent and one dependent variable are considered and the mathematical relationship between the two variables is linear.

Judging from the biological research literature, there seems to be a good deal of confusion about when one should use correlation and when one should use regression analysis. Some statisticians make a fairly rigid and clear-cut distinction between these two techniques, particularly with regard to the types of data to which they may be applied, while other statisticians, and apparently many biologists, seem to take a much more relaxed attitude about the whole affair and use the two approaches almost interchangeably. Examples 12.1, 12.2, and 12.3 will help to clarify the distinction between correlation and regression.

### • Example 12.1
A correlation problem

Data from a random sample of 11 gravid female iguanas, including their postpartum weights and the number of eggs each produced, were collected. The results are given in table 12.1, and the scatterplot ($x$-$y$ plot) for these data is shown in figure 12.1. (Notice that a line is *not* drawn through these points.)

We arbitrarily designate weight as the $x$ variable and the number of eggs as the $y$ variable. In a correlation analysis, it makes no difference which variable is plotted on which axis, as long as we are consistent from one observation to the next.

**Table 12.1** Postpartum weight of female iguanas and number of eggs produced

| Specimen Number | Mass (in Kilograms) | Number of Eggs |
|---|---|---|
| 1 | 0.90 | 33 |
| 2 | 1.55 | 50 |
| 3 | 1.30 | 46 |
| 4 | 1.00 | 33 |
| 5 | 1.55 | 53 |
| 6 | 1.80 | 57 |
| 7 | 1.50 | 44 |
| 8 | 1.05 | 31 |
| 9 | 1.70 | 60 |
| 10 | 1.20 | 40 |
| 11 | 1.45 | 50 |

Data from T. Miller

A cursory examination of the data in figure 12.1 leads us to suspect that larger females produce more eggs (i.e., that there is a relationship, or correlation, between a female's size and the number of eggs she produces in a single brood). However, we should be careful with our interpretation. The number of eggs a female produces is not the "cause" of her size, and conversely, her size is not necessarily a "cause" of the number of eggs she produces (although size might be an important contributing factor in determining the number of eggs). Both variables may, in fact, be under the control of a third variable, such as age or nutrition. If well-fed females tend to be large because of the amount of food they obtain, and if they also produce more eggs because of the amount of food they obtain, then size and number of eggs produced by a female are related, but both variables are under the control of a third variable: nutrition, in this case.

**Figure 12.1** Postpartum weight of female iguanas vs. number of eggs produced (example 12.1)

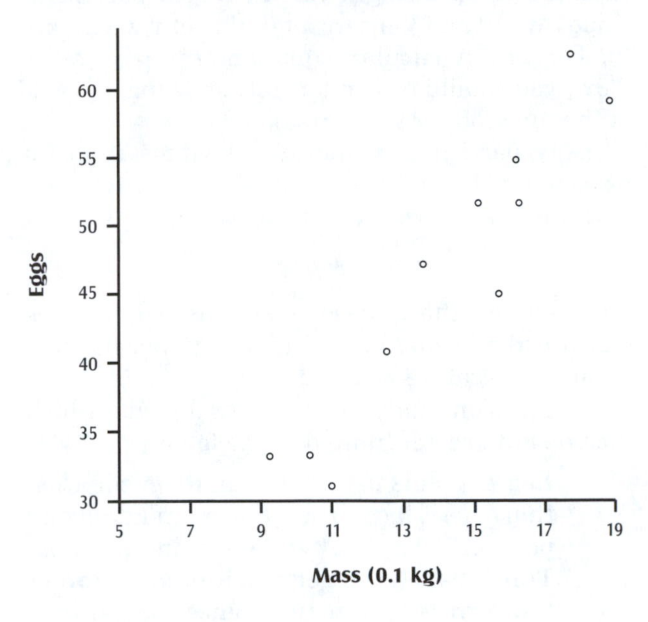


## • Example 12.2

### A regression problem

A snake physiologist wished to investigate the effect of temperature on the heart rate of juniper pythons. She selected nine specimens of approximately the same age, size, and sex and placed each animal at a preselected temperature between 2° and 18°C. After the snakes equilibrated to their ambient temperatures, she measured their heart rates. The results are given in table 12.2.

Note that, in this experiment, the temperatures were selected or under the control of the investigator. In this case temperature is not a random variable in the population.

**Table 12.2** The relationship between temperature and heart rate in juniper pythons

| Snake Number | Temperature (°C) | Heart Rate (BPM) |
|---|---|---|
| 1 | 2 | 5 |
| 2 | 4 | 11 |
| 3 | 6 | 11 |
| 4 | 8 | 14 |
| 5 | 10 | 22 |
| 6 | 12 | 23 |
| 7 | 14 | 32 |
| 8 | 16 | 29 |
| 9 | 18 | 32 |

The scatterplot for these data is shown in figure 12.2. Note that temperature is plotted on the $x$-axis, and heart rate is plotted on the $y$-axis, and not vice versa. Furthermore, we have drawn a line through the points and could, if we choose, write an equation that describes this line. Using this equation we could even predict the heart rate of juniper pythons at various temperatures.

**Figure 12.2** Heartbeat rate (BPM) of juniper pythons as a function of temperature (example 12.2)

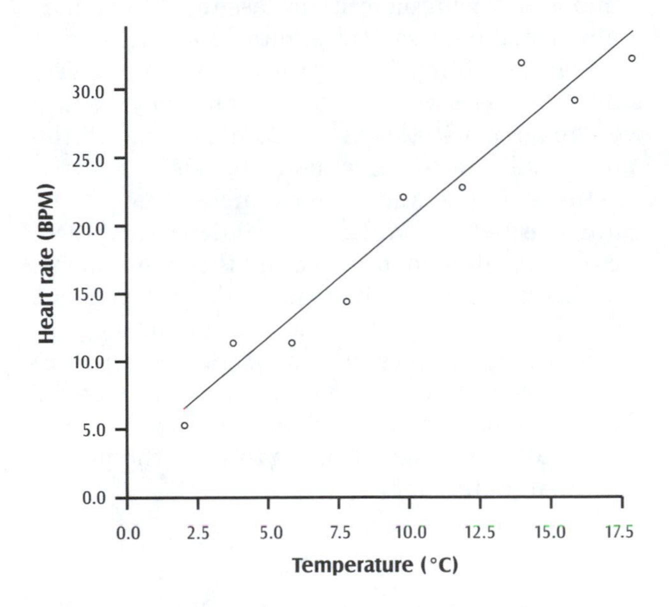


## • Example 12.3

### Another regression problem

Suppose in the study described in example 12.1 that the investigator had selected a sample of female iguanas based on their weight and determined the number of eggs that they produced. In this case the individuals included in the sample were not chosen at random but rather on the basis of the variable assumed to be the independent variable (i.e., the independent variable is under the control of the investigator in this case). Provided that we are willing to assume that the number of eggs that a female produces is at least partially "caused" or influenced by her weight, and that we can write an equation that will allow a prediction of the number of eggs a female might produce based on her weight, we would treat this situation as a regression problem.

To some extent the procedure used depends on the question being asked, and it always depends on the design of the experiment. In example 12.1, which is an example of a situation applicable to the correlation model, female iguanas were sampled at random. In example 12.3

they were selected on the basis of their size, and, therefore, size is not a random variable.

The first example is a proper problem for correlation analysis, while the second and third examples are better suited for regression analysis. While the two examples appear to be very much alike and to ask the same basic question (i.e., is there a relationship between the two variables?), there are many important differences between them. In the first example, gravid female iguanas were selected at random from a population. The two variables measured (postpartum weight and number of eggs produced) are assumed to be normally distributed variables in the population, and neither was controlled by the investigator. With data like these, we use correlation analysis, and we should resist the temptation to conduct the more sophisticated regression analysis!

In the second and third examples, the investigator selected the values of the independent (predictor) variable to be used in the experiment, and, therefore, in neither case is the independent variable treated as a random variable in the population. In both cases the dependent (response) variable is assumed to be at least partially under the control of the independent variable. In situations like these, regression analysis is the proper treatment of the data.

## 12.2 Simple Linear Regression Fundamentals

In regression analysis we assume a cause-and-effect relationship between the two variables under study. Furthermore, in most cases we also assume that the independent variable is under the control of the investigator (i.e., that a fixed-effects experimental design is used). In effect, we hypothesize that there is a functional relationship that permits us to predict a value of the dependent variable, $y$, corresponding to a given value of the independent variable, $x$. Mathematically, such a relationship is expressed as:

$$y = f(x)$$

In simple linear regression, the functional relationship between $y$ and $x$ takes the form

$$\mu_y = \alpha + \beta x \qquad (12.1)$$

where $\mu_y$ is the population mean value of $y$ at any value of $x$, $\alpha$ is the population intercept and $\beta$ is the population slope. Notice that this equation describes a straight line. Recall other similar formulas you may have seen (e.g., $y = a + bx$ or $y = mx + b$); these are all different ways of saying the same thing—a straight line having a particular slope ($\Delta y/\Delta x$) and $y$-intercept (value of $y$ where $x = 0$). For any particular equation (say $y = 2.0 + 0.8x$), you should be comfortable showing a graph of the function (or vice versa).

Any particular value of $y$ deviates from its expected value ($\mu_y$) due to some unexplained variation, which we call a **residual** ($e$):

$$y_i = \alpha + \beta x_i + e \qquad (12.2)$$

We assume that these residuals (sometimes called error terms) have a standard normal distribution (recall section 6.2).

Regression analysis has several goals, which include but are not limited to the following.

1. Regression is used to estimate an equation that describes the linear relationship between the two variables in question. This is called the **regression equation** or the regression function. Since the parameters $\alpha$ and $\beta$ are usually unknown to us, we estimate these values from a sample.
2. From this equation we are able to construct a line through the points of a scatterplot, which is called the **least squares regression line**.
3. The regression equation may be used to predict values of the dependent variable ($y$) at various values of the independent variable ($x$).
4. Regression may be used to estimate the extent to which the dependent variable is under the control of the independent variable.

Regression analysis is based on several assumptions. Unless these assumptions are met, one should either use correlation analysis or redesign the experiment. Sometimes, transformations may be used to correct certain problems in the data (recall section 10.5).

Some of these assumptions are rather obscure, but their meanings should become clear as we proceed. When assumption 1 is not met, regression is not the proper treatment for the data, but correlation might be.

Consider the imaginary experiment to determine the relationship, if any, between temperature and heart rate in juniper pythons (example 12.2). The investigator selected a series of temperatures (assumption 1) and designated this as the independent variable. She then placed a different animal at each of the preselected tempera-

### Assumptions of the Test

1. The independent variable is fixed. This means, in effect, that values of the independent variable are chosen by the investigator and do not represent a random variable in the population. There is thus no variance associated with the independent variable. (One kind of regression analysis, called "model II regression" and not covered in this book, does not require this assumption. For more information on model II regression, see Neter et al. (1989).)
2. For any value of the independent variable ($x$), there exists a normally distributed population of values of the dependent variable, $y$. The population mean of these values, $\mu_y$, is

   $$\mu_y = \alpha + \beta x$$

   where $\alpha$ is the population intercept and $b$ is the population slope of the regression equation.
3. The variances of the residuals for all values of $x$ are equal. (This is analogous to the homogeneous variance assumption of ANOVA.)
4. Observations are independent. In a practical sense, this means that each individual in the sample is measured only once.
5. The functional relationship is linear. To check this assumption, *always plot the data.*

tures and measured its heart rate (the dependent variable). The data obtained in the experiment are shown in table 12.3.

**Table 12.3** The effect of temperature on the heart rate of juniper pythons

| Temperature (°C) ($x$) | Heart Rate (BPM) ($y$) |
|---|---|
| 2 | 5 |
| 4 | 11 |
| 6 | 11 |
| 8 | 14 |
| 10 | 22 |
| 12 | 23 |
| 14 | 32 |
| 16 | 29 |
| 18 | 32 |
| $\Sigma x = 90$ | $\Sigma y = 179$ |
| $\bar{x} = 10$ | $\bar{y} = 19.88$ |

Note that the investigator measured only one individual at each temperature and that at each temperature a different individual was used (assumption 3). Had she measured the heart rate of one individual at different temperatures, the observations would not have been independent and any inferences about a relationship between the two variables would apply only to that single individual (a trivial result).

These data are shown as a scatterplot ($x$-$y$ plot) in figure 12.2 above. Note that although not all of the points seem to fall on a straight line, there seems to be a definite linear relationship between the two variables (assumption 5). Regression analysis can tell us more about this relationship.

## 12.3 Estimating the Regression Function and the Regression Line

When doing regression analysis, we are interested in estimating the parametric regression function (regression equation), $\mu_y = \alpha + \beta x$, where $\alpha$ is the intercept and $\beta$ is the slope of the equation. We estimate these values using a sample. The estimated intercept is designated as $a$ and the estimated slope, usually called the **regression coefficient**, is designated as $b$. The line described by this equation is the line that best "fits" the regression function, and it is called the estimated regression line. Since the line and the equation are one and the same (the equation defines the line), we consider them together in the following few sections. There is one point through which the regression line always passes, and that is the point defined by the mean of $x$ and the mean of $y$ ($\bar{x}$, $\bar{y}$). In figure 12.3 a horizontal line has been drawn through this point. A vertical line has been constructed from each value of $y$ to this horizontal line. Each of these vertical lines represents the amount by which the observed value of $y$ deviates from the mean value of $y$, or

$$y - \bar{y}$$

The sum of these $y \quad \bar{y}$ values is approximately 0 (table 12.4). However, if we square each of these $y \quad \bar{y}$ and then sum them, or

$$\Sigma(y - \bar{y})^2$$

we would have a sum of squares for $y$ (recall section 4.3.3). Note that this value is quite large (table 12.4).

**Figure 12.3** Graphic representation of dependent-variable variation when the independent variable is not considered (example 12.2)

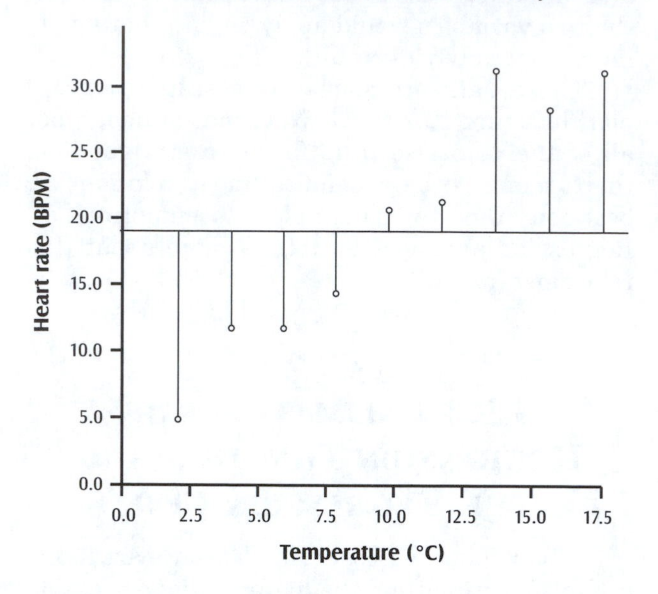


**Table 12.4** Computation of deviations and sum of squared deviations

| $x$ | $y$ | $y-\bar{y}$ | $(y-\bar{y})^2$ | $\hat{y}$ | $y-\hat{y}$ |
|---|---|---|---|---|---|
| 2 | 5 | −14.89 | 221.71 | 5.69 | −0.69 |
| 4 | 11 | −8.89 | 79.03 | 9.24 | 1.76 |
| 6 | 11 | −8.89 | 79.03 | 12.79 | −1.79 |
| 8 | 14 | −5.89 | 34.69 | 16.33 | −2.34 |
| 10 | 22 | 2.11 | 4.45 | 19.89 | 2.11 |
| 12 | 23 | 3.11 | 9.67 | 23.44 | −0.44 |
| 14 | 32 | 12.11 | 146.65 | 26.99 | 5.01 |
| 16 | 29 | 9.11 | 82.99 | 30.54 | −1.54 |
| 18 | 32 | 12.11 | 146.65 | 34.09 | −2.09 |
| Sums: 90 | 179 | −0.01 | 804.87 | $\Sigma(y-\hat{y})^2 = 48.74$ | |
| Means: 10 | 19.88 | | | | |

Data from table 12.3

Before terminal confusion sets in, we will review what we have just done here. In effect, we have calculated a sum of squares for $y$ without taking $x$ (temperature) into account. We can see that there is a great deal of variance in heart rate among our nine subjects when we do not consider their temperature.

Suppose now that we could rotate the line in figure 12.3, using $(\bar{x}, \bar{y})$ as a pivot, until it is in a position such that the deviations of the $y$ values from this line were minimized, or more specifically, in a position such that the sum of the squares of the deviations of the $y$ values from the line were minimized (figure 12.4). Such a line would best "fit" our data. (By the way, this is where the term "**least squares**" regression comes from!). You should note that the observed values of $y$ do not all fall on this line (sometimes none of them do). This is the regression line:

$$\hat{y} = a + bx \tag{12.3}$$

**Figure 12.4** Graphic representation of dependent-variable variation when the independent variable is considered (example 12.2)

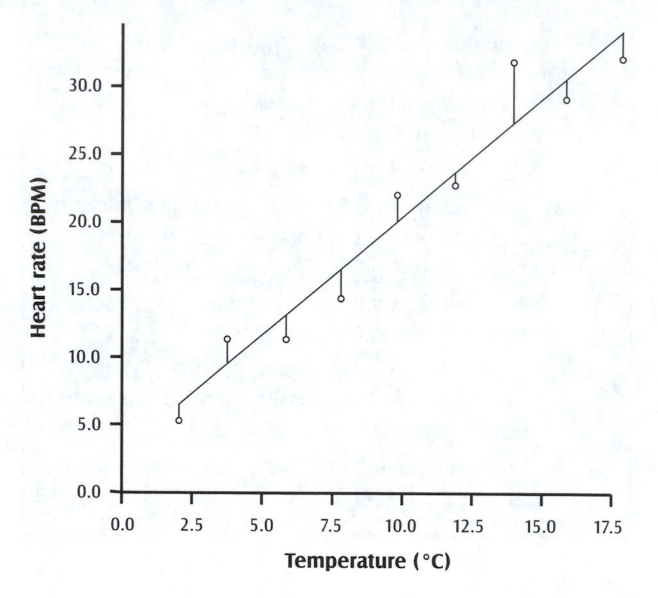


The value of $y$ that falls exactly on this line, as described by the regression equation, is referred to as $\hat{y}$ ("$y$ hat"). As before, we have constructed vertical lines from each value of $y$ to this new line, and these vertical lines denote the residuals: $(e = y - \hat{y})$. These values, given in table 12.4, represent the amount by which each observed value of $y$ deviates from the regression line. For the moment do not be concerned with how we defined the regression line or how the values of $\hat{y}$ were calculated; we return to this later. Squaring the values of $y - \hat{y}$ in table 12.4 and then summing them gives us another sum of squares for $y$. This sum of squares represents the variation in $y$ when temperature ($x$) is taken into account. Note that it is much smaller than the variation in $y$ when $x$ was not considered. In statistical jargon, we have decreased the uncertainty of $y$ by considering $x$. What does this mean exactly?

If we have information regarding the heartbeat rate of nine juniper pythons (example 12.2) but no knowledge of their temperatures, and if from these data we need to predict something about the heartbeat rate of juniper pythons generally, we could only conclude that the mean heartbeat is around 19.89 BPM with a great deal of variation from animal to animal (estimated from $\Sigma(y-\bar{y})^2$, which is large). On the other hand, given a knowledge of $x$ (temperature), we could make a much more accurate prediction of the heartbeat rate of juniper pythons.

Refer once again to table 12.4 and figure 12.4. Note that the values of $y$ do not exactly coincide with the values of $\hat{y}$. The amount by which $y$ and $\hat{y}$ differ $(y-\hat{y})$ is called a **residual** and is denoted by the symbol $e$. One of the assumptions of regression is that the residuals are normally distributed with a mean of zero (assumption 2, in part).

## 12.4 Calculating the Estimated Regression Equation

Now we are ready for some formulas, used to estimate the regression parameters. The esti-

$$b = \frac{\Sigma xy - \frac{\Sigma x \Sigma y}{n}}{\Sigma x^2 - \frac{(\Sigma x)^2}{n}} \tag{12.4}$$

mated **slope** of the regression equation is given by To obtain this equation, statisticians must solve some simultaneous equations. We'll leave out those details here, although interested readers may consult Neter et al. (1989). Since the least squares regression line goes through the point $(\bar{x}, \bar{y})$, the estimated **y-intercept** is given by

$$a = \bar{y} - b\bar{x} \tag{12.5}$$

**Table 12.5** Intermediate calculations for regression statistics

| | $x$ | $y$ | $xy$ | $x^2$ | $y^2$ |
|---|---|---|---|---|---|
| | 2 | 5 | 10 | 4 | 25 |
| | 4 | 11 | 44 | 16 | 121 |
| | 6 | 11 | 66 | 36 | 121 |
| | 8 | 14 | 112 | 64 | 196 |
| | 10 | 22 | 220 | 100 | 484 |
| | 12 | 23 | 276 | 144 | 529 |
| | 14 | 32 | 448 | 196 | 1024 |
| | 16 | 29 | 464 | 256 | 841 |
| | 18 | 32 | 576 | 324 | 1024 |
| Sums | 90 | 179 | 2216 | 1140 | 4365 |
| | $\Sigma x$ | $\Sigma y$ | $\Sigma xy$ | $\Sigma x^2$ | $\Sigma y^2$ |

Data from table 12.3

For the juniper python example (table 12.5),

$$b = \frac{2216 - \frac{90 \times 179}{9}}{1140 - \frac{(90)^2}{9}} = 1.775$$

and

$$a = 19.89 - (1.775 \times 10) = 2.14$$

The estimated regression function is therefore

$$\hat{y} = 2.14 + 1.775x$$

The regression line is defined by the values of $\hat{y}$ corresponding to values of $x$. These values are shown in table 12.6. The regression line may be constructed graphically by plotting the values of $\hat{y}$ versus the corresponding values of $x$. We have now obtained estimates of the slope of the regression equation ($b$), its $y$-intercept ($a$), and the regression line. We now need to attach some statistical significance to these estimates. Table 12.6 provides the intermediate statistics needed for these statistical tests. Work through the math with your calculator and/or spreadsheet to verify these numbers.

**Table 12.6** The effect of temperature on the heart rate of juniper pythons

| Temperature (°C) ($x$) | Heart rate ($y$) | $\hat{y}$ | $e$ (residual) |
|---|---|---|---|
| 2 | 5 | 5.69 | −0.69 |
| 4 | 11 | 9.24 | 1.76 |
| 6 | 11 | 12.79 | −1.79 |
| 8 | 14 | 16.33 | −2.34 |
| 10 | 22 | 19.89 | 2.11 |
| 12 | 23 | 23.44 | −0.44 |
| 14 | 32 | 26.99 | 5.01 |
| 16 | 29 | 30.54 | −1.54 |
| 18 | 32 | 34.09 | −2.09 |
| $\Sigma x = 90$ | $\Sigma y = 179$ | $\Sigma(y-\bar{y})^2 = 804.89$ | |
| $\bar{x} = 10$ | $\bar{y} = 19.89$ | $\Sigma(y-\hat{y})^2 = 48.74$ | |
| $\Sigma x^2 = 1140$ | $\Sigma y^2 = 4365$ | | |
| $(\Sigma x)^2 = 8100$ | $(\Sigma y)^2 = 32041$ | | |
| $\Sigma xy = 2216$ | $\Sigma x \Sigma y = 16110$ | | |

## 12.5 Testing the Significance of the Regression Equation

Recall that the sample slope, $b$, is an estimate of the parametric slope, $\beta$. Even if $\beta = 0$, which would indicate no dependence of $y$ on $x$, we might expect $b$ to occasionally have a nonzero value by chance alone. We therefore test the null hypothesis $H_0$: $\beta = 0$. We could evaluate this null hypothesis either with ANOVA or a $t$-test. Here we use ANOVA for the hypothesis test. To conduct the ANOVA, we need three sums of squares and their associated degrees of freedom: the total sum of squares with $n - 1$ degrees of freedom, the regression sum of squares with 1 degree of freedom, and the error sum of squares with $n - 2$ degrees of freedom. Table 12.6 gives the values needed to calculate these sums of squares by the most direct methods, shown below. Alternative approaches, which give the same answers for $SS_t$ and $SS_e$, are shown in table 12.4. Check that table as you work through the example below to see that these answers are equivalent.

The total sum of squares, $\Sigma(y - \bar{y})^2$, is given by

$$SS_t = \Sigma y^2 - \frac{(\Sigma y)^2}{n} \qquad (12.6)$$

This is sum of squares for $y$ when $x$ is not considered. (Notice the parallel with equation 4.4 used earlier for descriptive statistics.) For the current example,

$$SS_t = 4365 - \frac{(179)^2}{9} = 804.89$$

The regression sum of squares is given by

$$SS_r = b \times \left( \Sigma xy - \frac{\Sigma x \Sigma y}{n} \right) \qquad (12.7)$$

This is the sum of squares for $y$ when $x$ is considered.

For the example,

$$SS_r = 1.775 \times \left( 2216 - \frac{90 \times 179}{9} \right) = 756.15$$

The error sum of squares is given by

$$SS_e = SS_t - SS_r \qquad (12.8)$$

This is the sum of squares of the residuals, and it contributes to the variance in $y$ that is still present when $x$ is considered. In other words, it is the error sum of squares. Refer back to figure 12.4, which illustrates these residuals as deviations of points from the regression line.

For the example,

$$SS_e = 804.89 - 756.15 = 48.74$$

The ANOVA table is constructed as before (section 10.3).

**Table 12.7** ANOVA table for the data in table 12.3

| Source | SS | df | MS | F |
|---|---|---|---|---|
| Regression | 756.15 | 1 | 756.15 | 108.60 |
| Error | 48.74 | 7 | 6.96 | |
| Total | 804.89 | 8 | | |

From table A.6, the critical value of $F$ for 1 and 7 df at $\alpha = 0.05$ ($F_{1,7,(0.05)}$) is 5.59. Since 108.60 >> 5.59, we reject $H_0$: $\beta = 0$ and conclude that the value of $y$ is dependent on the value of $x$ (i.e., that the regression slope is not zero). The very large $F$ statistic indicates that the $p$ value is quite small.

Next let's see how MINITAB displays these same statistics (table 12.8). Notice that the regression equation is the same as we calculated earlier (section 12.4). Below the equation are statistics relating to the $y$-intercept and the slope. MINITAB displays a $t$ statistic that evaluates $H_0$: $\beta = 0$ (no dependence of $y$ on $x$). The very small $p$ value indicates this hypothesis is rejected (i.e., the dependence of $y$ on $x$ is very strong). We already saw that dependence visually in the scatterplot (figure 12.2). The strength of the dependence is reported as "R-Sq" (more about this in section 12.7). The ANOVA table matches what we calculated in table 12.7. MINITAB also reports a very small $p$ value (<0.001). "Unusual observations" lists the seventh observation, which has a large residual (see figure 12.4).

**Table 12.8** MINITAB output for simple linear regression of heart rate by temperature for the data in table 12.3

**Regression Analysis: HeartRate versus Temperature**

```
The regression equation is
HeartRate = 2.14 + 1.77 Temperature

Predictor        Coef   SE Coef      T        P
Constant        2.139     1.917   1.12    0.301
Temperature    1.7750    0.1703  10.42    0.000

S = 2.63869    R-Sq = 93.9%     R-Sq(adj) = 93.1%

Analysis of Variance
Source          DF       SS        MS        F        P
Regression       1   756.15    756.15   108.60    0.000
Residual Error   7    48.74      6.96
Total            8   804.89

Unusual Observations

 Obs   Temperature   HeartRate     Fit   SE Fit   Residual   St Resid
   7          14.0      32.000  26.989    1.113      5.011      2.09R

R denotes an observation with a large standardized residual.
```

## 12.6 The Confidence Interval for $\beta$

Recall that $b$ is an estimate of $\beta$, the regression coefficient (the true population slope). As we have seen when estimating other population parameters, we cannot say with certainty what the exact value of $\beta$ is; however we can compute a 95% confidence interval for $\beta$. This requires first determining the amount of uncertainty in our estimate of the slope. The standard error for the slope, $s_b$, is given by

$$s_b = \sqrt{\frac{\text{MS}_e}{\sum x^2 - \frac{[\sum x]^2}{n}}} \qquad (12.9)$$

where $\text{MS}_e$ is the mean square for error (from the ANOVA table) and the denominator is the sum of squares for $x$ (equation 4.4). The 95% confidence interval for $\beta$ is then

$$b \pm s_b\,(t_{0.05,\, n-2}) \qquad (12.10)$$

where $t$ is the two-tailed probability of $t$ at alpha = 0.05 with $n - 2$ degrees of freedom. For our example, $t_{7,0.05} = 2.365$. Thus,

$$s_b = \sqrt{\frac{6.96}{1140 - \frac{8100}{9}}} = 0.1703$$

Notice that this statistic is also reported by MINITAB above (table 12.8). Then, the 95% confidence interval for $\beta$ is:

$$1.775 \pm (0.1703 \times 2.365)$$
$$= 1.775 \pm 0.403$$

The lower and upper confidence interval for $\beta$ is thus 1.372 to 2.178.

## 12.7 The Coefficient of Determination ($r^2$)

We know that the variance in $y$ is greatly reduced by a knowledge of $x$, but there is usually still some variance remaining in $y$ when $x$ has been considered (the error variance). If the value of $y$ were completely dependent on $x$, there would be no error variance, which is to say that all of our observations would fall on the regression line. So what proportion of the variance in $y$ is explained by its dependence on $x$? To determine this, we compute the **coefficient of determination**, $r^2$. This value is computed by

$$r^2 = \frac{\text{SS}_r}{\text{SS}_t} \qquad (12.11)$$

For the example,

$$r^2 = \frac{756.15}{804.89} = 0.939$$

Thus, we may conclude that 0.939 or 93.9% of the variance in $y$ is dependent on $x$, which is to say that when we know the value of $x$, we reduce the uncertainty about $y$ by 93.9%. There is still a residual or "unexplained" variance of 100 – 93.9 = 6.1% of the variance in $y$ that is still not explained. This is the variance among individuals that is not related to $x$. If all the points fell exactly on a straight line, $r^2$= 100% and the unexplained variance would be zero.

The coefficient of determination may also be computed for a correlation analysis (chapter 13), in which case it is the square of the correlation coefficient, $r$. However, in correlation treatments, $r^2$ should not be considered as a measure of the variation in $y$ that is explained or dependent upon $x$, but rather as the variation in $y$ that is *associated* with the variance of $x$, and vice versa.

## 12.8 Predicting $y$ from $x$

One important use of regression is to enable us to predict a value of $y$ for a given value of $x$. Simply plug a number of $x$ into the equation or find the value of $y$ off the graph. Such predictions must be made with some restraint, however. In the juniper-python example, we measured heart rate ($y$) for temperatures ($x$) between 2°C and 18°C. Predictions about heart rate much beyond these measured values should be avoided, since we run the risk of predicting nonsense. For example, we would predict a very high heart rate at 100°C by using the regression function (over 179 BPM), but biologically we would probably predict a heart rate of zero (the snakes are dead!). In a similar manner we would probably predict a negative heart rate at temperatures much below 0°C, which is clearly impossible.

Even when predictions about $y$ are kept within reasonable limits of $x$, it is important to remember that $y$ is usually not an exact function of $x$, since $y$ is a normally distributed random variable. Thus, when we predict a value of $y$ from $x$, what we are in fact doing is estimating the population mean value of $y$ for any particular value of $x$. This estimated value is designated as $\hat{y}$. As usual, we should affix confidence limits to this estimate.

To continue with the juniper-python example, we wish to know the mean heart rate of all juniper pythons at 15°C. First, we calculate the value of $\hat{y}$ at $x$ = 15 degrees from the regression equation, which is

$$\hat{y} = 2.14 + (1.775 \times 15) = 28.77$$

**Caution**

Although in mathematics, a line extends forever in both directions, in statistics, the line (segment) should only extend over the range of the data (minimum $x$ to maximum $x$) and no further. Extending the line further in either direction is incorrect because we do not know if the functional relationship continues to be linear beyond the points we observed. Extending the regression line beyond the range of the points used to generate it is called **extrapolation**, which should usually be avoided. There are times when extrapolation is used, such as in toxicology research. However, interpretation can be very tricky and rests on making some assumptions. For an example, see figure 12.5. This figure illustrates a hypothetical toxicology experiment on mice, interpreted to doses to which humans are exposed in the environment. In order to see an effect with a reasonable number of mice (we only have room for so many cages!), the experiment must use high doses. Now, assuming that humans have the same sensitivity as mice and that the function is similar at low doses (continues to be linear), we estimate that a dose of 0.0002 ppm would lead to a cancer risk of 0.000022 (about 22 cases per million). Toxicologists argue a lot about these kinds of experiments.

**Figure 12.5** An example of extrapolation in toxicology

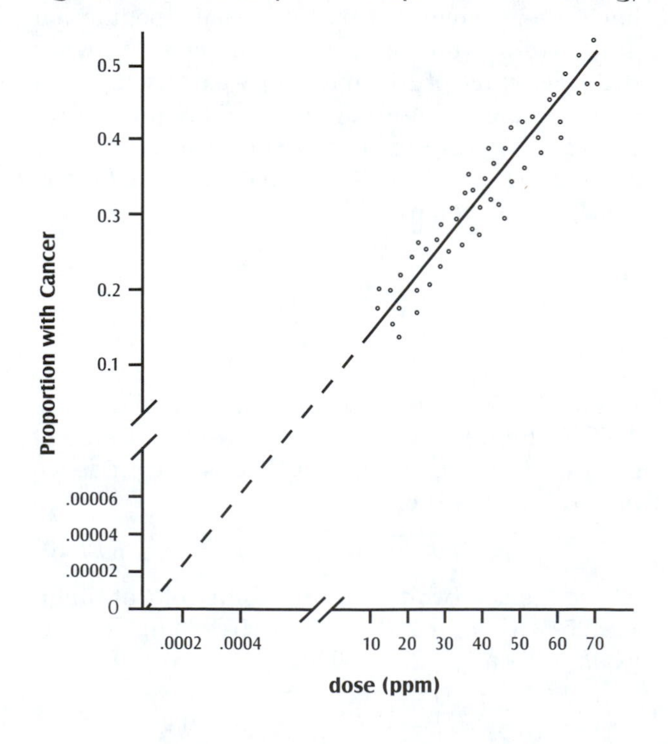

We now compute the standard error of $y$, which is given by

$$s_{\hat{y}} = \sqrt{MS_e\left[\frac{1}{n} + \frac{(x-\bar{x})^2}{\sum x^2 - \frac{(\sum x)^2}{n}}\right]} \quad (12.12)$$

where $x$ is that value of $x$ for which we want a confidence interval for $y$, and $\bar{x}$ is the mean value of $x$. For the example,

$$s_{\hat{y}} = \sqrt{6.96\left[\frac{1}{9} + \frac{(15-10)^2}{(1140 - \frac{90^2}{9})}\right]} = 1.224$$

The 95% confidence interval for the predicted mean value of $y$ ($\mu_{\hat{y}}$) is:

$$\hat{y} \pm (s_{\hat{y}}) \times t_{n-2,(0.05)} \quad (12.13)$$

where $t$ is the critical value of $t$ at alpha = 0.05 and $n - 2$ degrees of freedom (2.365). For the example, $t_{7,(0.05)} = 2.365$ and the 95% confidence interval for $\mu_{\hat{y}}$ is:

$$28.77 \pm (1.224 \times 2.365) \pm 2.895$$

In other words, the 95% confidence interval for $\mu_{\hat{y}}$ is 25.87 to 31.67.

We conclude that there is a probability of 0.95 that these limits include the population value of the true heart rate at 15°C. A graphic representation of the confidence interval (or prediction interval) of $\mu_y$ at various values of $x$ is shown in figure 12.6. The lines above and below the regression line show the 95% prediction interval for $\mu_y$ at any value of $x$ within our measured limits of $x$. Note that the prediction interval becomes wider at lower and higher values of $x$. In other words, as we get farther away from the mean of $x$ our uncertainty in estimating $y$ increases.

**Figure 12.6** The 95% confidence interval (prediction interval) for example 12.2

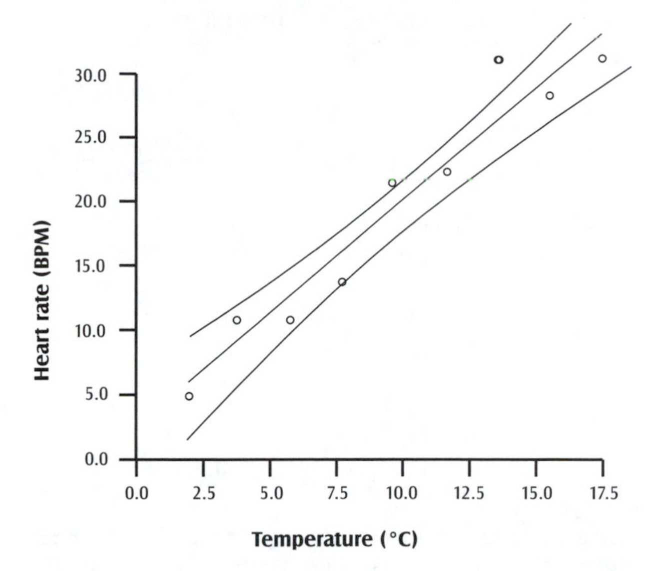


## 12.9 Dealing with Several Values of *Y* for Each Value of *X*

Frequently, data that are to be analyzed by regression are conducted as we have outlined previously. Note that in our example, there is only one value of $y$ for each value of $x$, which is to say that only one snake was measured at each temperature. For various reasons, it is sometimes desirable or necessary to measure the value of the dependent variable, $y$, for several individuals at each value of the independent variable, $x$. For example, an investigator might wish to test only a few values of $x$. Using only one measurement of $y$ at each $x$ would result in a sample size that might be too small to reveal a significant association (i.e., a sample size that would result in a high probability of a type II error).

### • Example 12.4

More than one value of *y* for each value of *x*

For purposes of illustration, we will redesign the juniper-python experiment somewhat, this time using only three temperatures but measuring the heart rate of several snakes at each temperature. The results of the experiment are given in table 12.9.

With the data arrayed in this way, we may proceed exactly as before when there was only one value of $y$ for each value of $x$. Note carefully, however, that the sum of $x$ is not 4 + 10 + 16; rather it is (4 × 4) + (6 × 10) + (5 × 16). In other words, we do not have three values of $x$—we have 15, some of which are the same! The scatterplot for these data, showing the regression line and the 95% prediction interval, is given in figure 12.7.

**Table 12.9** The effect of temperature on the heart rate of juniper pythons

| Temperature | Heart Rate |
|---|---|
| 4 | 9 |
| 4 | 8 |
| 4 | 11 |
| 4 | 8 |
| 10 | 20 |
| 10 | 21 |
| 10 | 19 |
| 10 | 20 |
| 10 | 19 |
| 10 | 20 |
| 16 | 30 |
| 16 | 28 |
| 16 | 31 |
| 16 | 29 |
| 16 | 30 |

**Figure 12.7** The 95% confidence interval for multiple measurements at each value of $x$ (example 12.4)

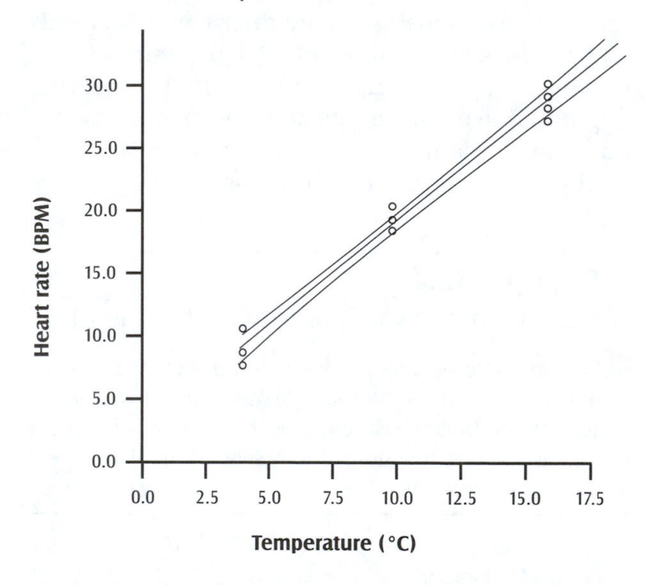


**Caution**

In the example we have just examined, 15 snakes were used. It is sometimes tempting to measure the same individual repeatedly at each value of $x$, which, in our example, would be only three snakes. This is a serious violation of the independence assumption (see chapter 1 and section 12.2 above). Had we done the experiment in this way, we could use the mean value of $y$ at each $x$ as a datum (which is sometimes useful), but our sample size would be 3, not 15.

## 12.10 Checking Assumptions and Remedies for Their Failure

As we saw above (section 12.2), for linear regression to work properly several assumptions must be met. The simplest check of these assumptions is to create a scatterplot of the data. This allows us to check for linearity (assumption 2) and also to check for large departures from equal variances (assumption 4). For instance, consider figure 12.8. Panel A illustrates a case where the relationship is strong but clearly nonlinear. Panel B shows a case where the error variance increases as $x$ increases. In both cases, use of linear regression would be inappropriate.

With statistical software at our disposal, other plots and more formal tests are available. For instance, an option with the MINITAB regression-analysis routine is to store the residuals. Residuals plots of $e$ against $x$ (or against $\hat{y}$) allow us to more clearly visualize the departure from equal variance. Residuals can be checked for normality (section 10.4). If we know the order of data collection, residuals can also be checked for independence by other procedures.

When the relationship between the variables in a regression analysis does not seem to be linear or the variances are not equal, it is sometimes possible to correct the data by transforming one or both of the variables by methods we saw earlier (see section 10.5).

## 12.11 Advanced Regression Techniques

In this chapter, you have been introduced to simple linear regression. Regression analysis is a versatile technique and, like ANOVA, has been extended to many different situations. For instance, other mathematical functions may be applied to nonlinear data (e.g., exponential, sigmoidal, etc.). Some statistical software can fit functions to these data directly. Alternatively, using the appropriate transformation (e.g. log function on exponential data) allows converting the data to a linear scale, after which linear regression may be performed on the transformed data.

Regression may also be used to model the dependence of $y$ on multiple predictor variables.

For instance, we might model the heart rate of our snakes to simultaneous changes in both temperature and exposure to some environmental chemical. Modelers sometimes call these "response surfaces," because graphs can be visualized in three dimensions.

Another widely used technique is comparing regression lines from two or more groups—a procedure called analysis of covariance (ANCOVA). A simple graphic example for two groups is shown in figure 12.9. ANCOVA allows us to ask three questions about the data:

(1) Does $y$ depend on $x$? (2) Is the mean of $y$ for one group different from other groups? (3) Is the response (slope) of $y$ against $x$ different between groups?

**Figure 12.8** Examples of failed regression assumptions

A

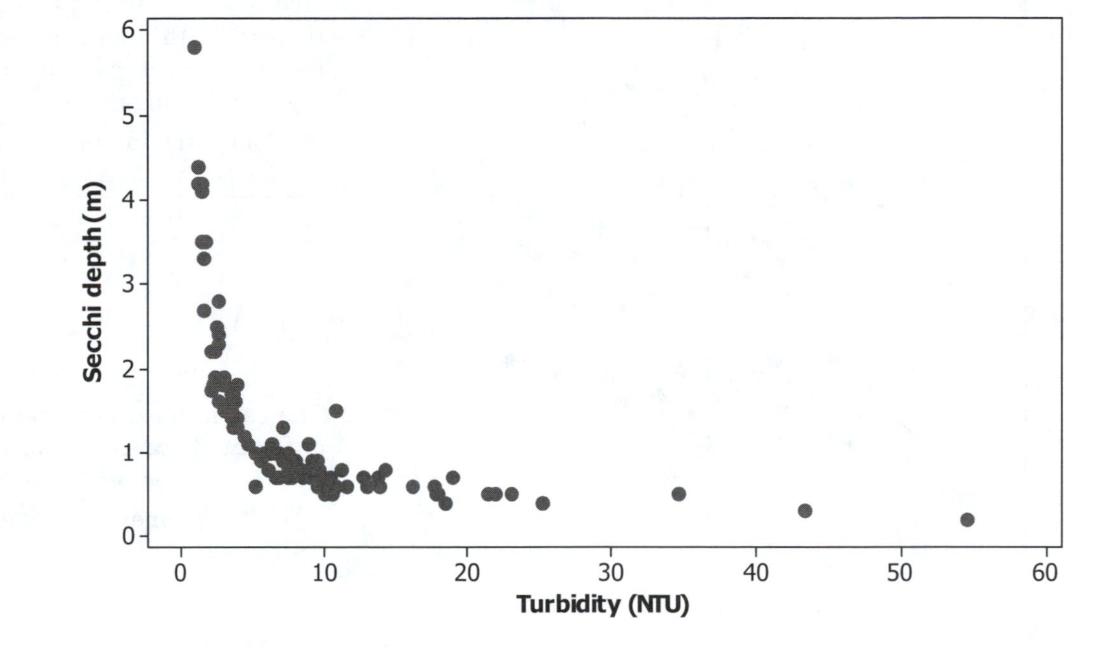


B

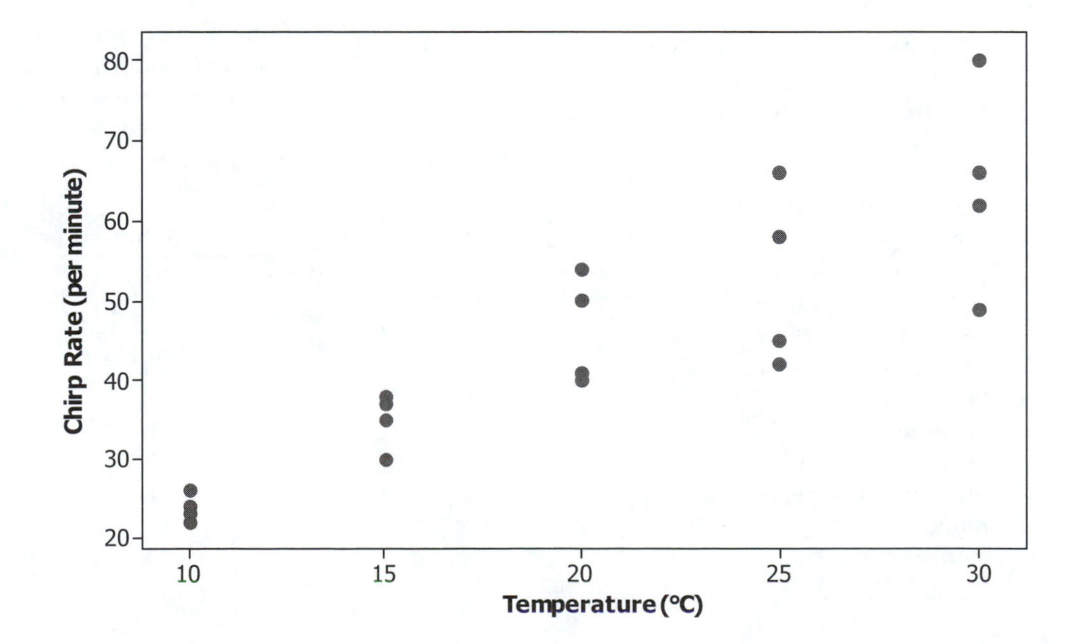

Students who expect to use regression analysis in their research are encouraged to explore more advanced texts, such as Neter et al. (1989), listed in the references (appendix C).

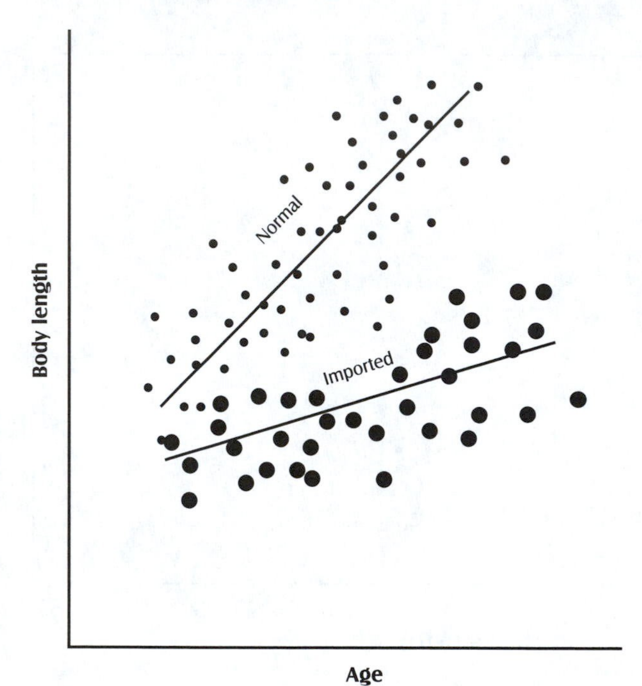


**Figure 12.9** Comparing regression lines. In this case, the different growth rates of the two populations are reflected by differences in the slope of their size-by-age regression lines

## Key Terms

coefficient of determination
correlation analysis
correlation coefficient
dependent variable
extrapolation
independent variable
least squares
predictor variable
regression analysis
regression coefficient
regression equation
regression line
residual
response variable
simple linear regression
slope
$y$-intercept

## Exercises

For each exercise, first graph and inspect a scatterplot of the two variables. Check for any obvious departures from the regression assumptions. Compute the regression equation and coefficient of determination. Test the null hypothesis that $\beta = 0$. Compute the 95% confidence interval for the slope and the 95% confidence interval for $\mu_y$ for each value of $x$. Answer any additional questions specific to each exercise.

12.1 Algae cells were incubated in a culture medium containing different concentrations of dilithium chloride. After a period of incubation, the concentration of dilithium ($\mu$g/g) in the algae cells was determined.

| Concentration in medium ($\mu$M/$L$) | Concentration in cells ($\mu$g/g) |
|---|---|
| 0 | 0 |
| 1 | 9 |
| 2 | 21 |
| 5 | 47 |
| 10 | 105 |
| 20 | 213 |

12.2 The cadmium concentrations of grasses at different distances from a major highway were measured, with the following results.

| Distance (meters) | Cadmium ($\mu$g/kg) |
|---|---|
| 1 | 105 |
| 2 | 48 |
| 3 | 39 |
| 4 | 28 |
| 5 | 18 |
| 10 | 9 |

12.3 9 male road warblers of different ages were selected and their systolic blood pressure was measured. (Nine or 10 years is considered very old among road warblers.) The results were as follows.

| Age (years) | Blood Pressure (mm Hg) |
|---|---|
| 1 | 103 |
| 2 | 115 |
| 3 | 109 |
| 4 | 114 |
| 5 | 120 |
| 6 | 119 |
| 7 | 128 |
| 8 | 132 |
| 9 | 138 |

12.4 The wattle thickness in chickens given 5 concentrations of PHA (and control) was as follows.

| PHA concentration ($\mu$g/L) | Wattle Thickness (mm) |
|---|---|
| 0 | 1.01 |
| 1 | 1.53 |
| 2 | 2.00 |
| 3 | 2.47 |
| 4 | 3.01 |
| 5 | 3.66 |

12.5 The rate of an enzyme catalyzed reaction (o-diphenol oxidase) was measured at different substrate concentrations. Rate is the dependent variable. A reciprocal transformation of both variables is required to make the relationship linear. Try plotting both the original data and the transformed data, and note the difference in appearance.

| Substrate Concentration (mM) | Rate ($\mu$L $O_2$/min) |
|---|---|
| 0.20 | 105.3 |
| 0.30 | 142.9 |
| 0.40 | 166.7 |
| 0.50 | 181.8 |
| 1.00 | 256.4 |

12.6 10 colonies of juvenile hamsters were established with different densities of animals ranging from 1–5 animals per square meter. After one week, three animals were randomly selected from each colony and their serum corticosterone was measured. The relationship does not appear to be linear. You might wish to try one or more transformations (see section 10.5) to attempt to rectify the situation.

| Density | Serum Corticosterone |
|---|---|
| 1 | 3.2 |
| 1 | 2.8 |
| 1 | 3.1 |
| 2 | 8.5 |
| 2 | 10.2 |
| 2 | 9.9 |
| 3 | 27.5 |
| 3 | 34.0 |
| 3 | 29.8 |
| 4 | 97.2 |
| 4 | 120.0 |
| 4 | 105.6 |
| 5 | 330.0 |
| 5 | 285.0 |
| 5 | 315.5 |

12.7 Tomato plants of the same genetic strain and age were subjected to a temperature of 115°F for a period of three hours. Such treatment reduces the plants' ability to photosynthesize. One plant from the group was randomly selected each day for 11 consecutive days, and its rate of photosynthesis was determined. We wish to know if the plants recover from this temperature stress, and if so, write an equation to describe the rate of recovery.

| Days Posttreatment | Photosynthetic Rate ($\mu$M $CO_2$/gram/sec) |
|---|---|
| 0 | 15.0 |
| 1 | 17.5 |
| 2 | 16.5 |
| 3 | 19.0 |
| 4 | 22.0 |
| 5 | 24.0 |
| 6 | 22.5 |
| 7 | 26.5 |
| 8 | 25.0 |
| 9 | 30.0 |
| 10 | 29.0 |

12.8 A certain species of bacterium was cultured in the presence of various concentrations of methyl mercury. The number of cells (per milliliter) in the various cultures was determined after a period of incubation.

| Methyl Mercury ($\mu$M) | Cells $\times$ $10^6$/ml |
|---|---|
| 0 | 6.6 |
| 0 | 6.9 |
| 0 | 7.2 |
| 1 | 6.8 |
| 1 | 6.0 |
| 1 | 5.6 |
| 2 | 6.4 |
| 2 | 6.0 |
| 2 | 5.4 |
| 4 | 4.8 |
| 4 | 4.4 |
| 4 | 3.9 |
| 6 | 2.6 |
| 6 | 3.1 |
| 6 | 3.4 |
| 8 | 1.0 |
| 8 | 1.3 |
| 8 | 1.7 |
| 10 | 0.2 |
| 10 | 0.3 |
| 10 | 0.5 |

12.9 Kingfisher nestlings 5–7 days old were incubated at different ambient temperatures, and their body temperatures were measured. At this age the birds are not capable of maintaining a constant body temperature as are older birds. The nestlings were randomly assigned to the various temperatures. (Data from M. Hamas.)

| Ambient Temperature (°C) | Body Temperature (°C) |
|---|---|
| 10 | 10 |
| 10 | 12 |
| 10 | 11 |
| 10 | 17 |
| 10 | 15 |
| 10 | 13 |
| 20 | 21 |
| 20 | 28 |
| 20 | 27 |
| 20 | 27 |
| 20 | 24 |
| 20 | 25 |
| 20 | 20 |
| 30 | 29 |
| 30 | 31 |
| 30 | 28 |
| 30 | 35 |
| 30 | 36 |
| 30 | 30 |
| 40 | 40 |
| 40 | 39 |
| 40 | 35 |
| 40 | 37 |
| 40 | 37 |
| 40 | 40 |
| 40 | 38 |

# 13

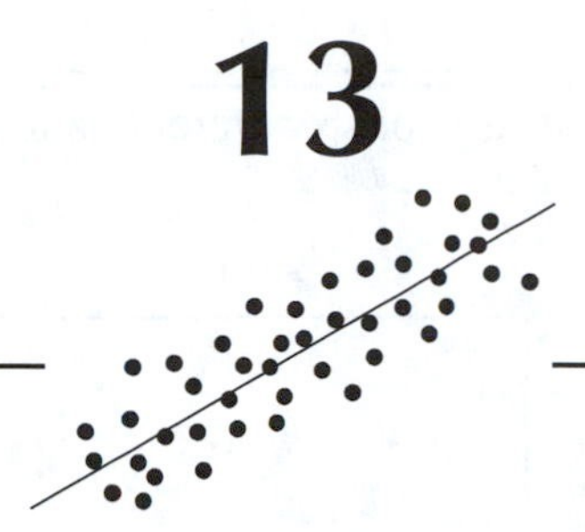

# Association between Two Measurement Variables

## Correlation

We are often interested in knowing whether two variables tend to be associated with each other. Do two methods of measuring blood pressure tend to give similar results? How strongly associated are pairs of morphometric characteristics of grizzly bears? Is there a correspondence between concentrations of two toxic metals (e.g., cadmium and lead) in the sediments of streams in a watershed impacted by industrial pollution? These questions all ask for correlation analysis. In this chapter we are solely interested in determining the degree to which two measurement variables are associated. In the next chapter, we explore a similar question, but look at categorical variables (section 14.2).

In correlation analysis, we ask two questions: (1) Are two measurement variables related in some consistent and linear way and, if so, in what direction? And (2) What is the strength of that relationship? The strength of such relationships is represented graphically by how closely the points in a scatterplot of the two variables cluster about an imaginary line drawn through them. (Note that we do not actually draw this line, and it would be improper to do so in a correlation analysis.)

### 13.1 The Pearson Correlation Coefficient

The measure of the strength of the relationship between two variables in a correlation is the **correlation coefficient**, formally called the Pearson correlation coefficient or the product moment correlation coefficient. The correlation coefficient of the population is designated by the Greek symbol rho ($\rho$). Usually, the true value of this parameter is unknown to us, and we must estimate its value from a random sample of the population. The sample correlation coefficient is designated as $r$ The value of $\rho$ (or $r$) ranges from +1, indicating a perfect positive correlation (all points falling on one line with a positive slope); through 0, indicating no relationship between the two variables; to –1, indicating a perfect negative correlation between the two variables. See figure 13.1, which illustrates different degrees of association between pairs of variables. Note that we are not saying anything about the value of the slope. In fact, with correlation, slope is entirely irrelevant to the question. We only want to answer questions 1 and 2 above.

**Figure 13.1** Scatterplots and correlation statistics for six different data sets showing different degrees of association between two variables

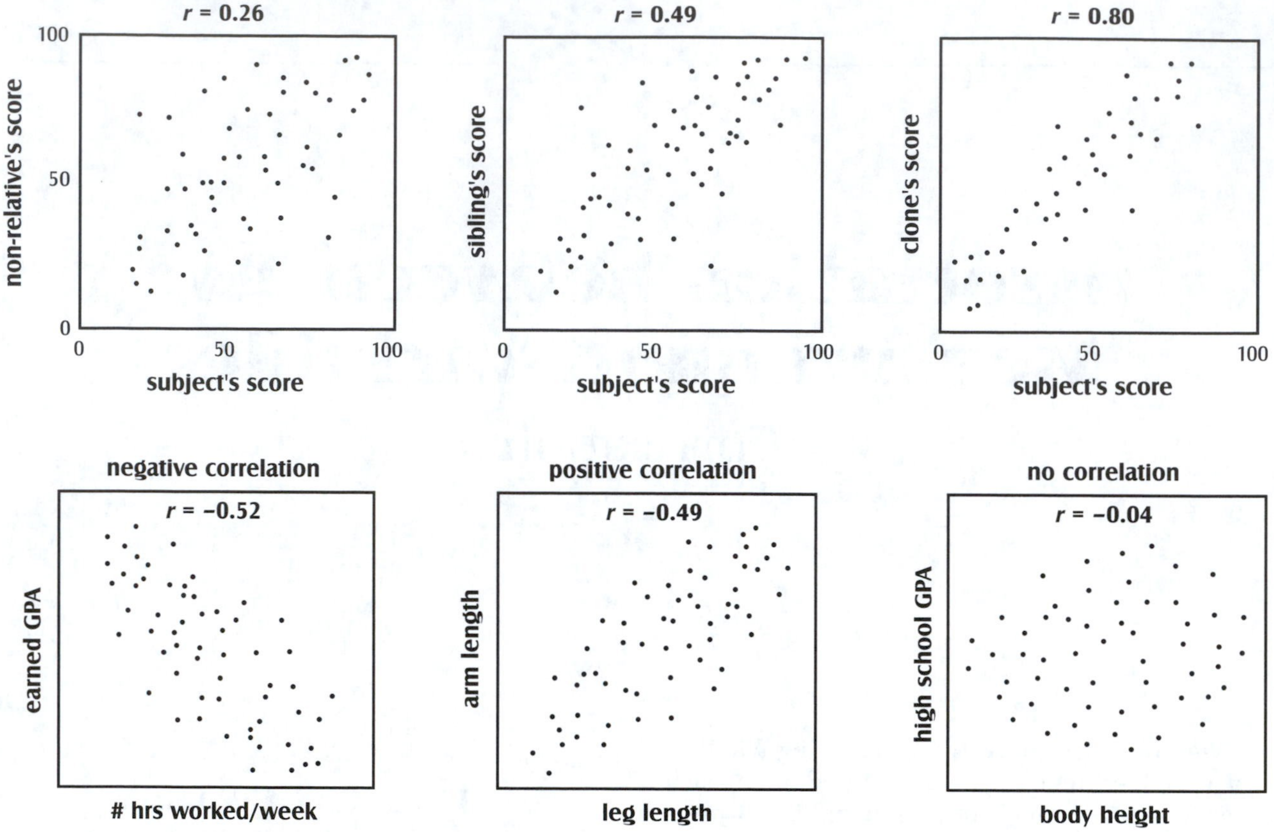

Adapted from J. Sumich, unpublished.

The null hypothesis in a correlation analysis is that the parametric correlation coefficient, $\rho$, is zero, or

$$H_0: \rho = 0$$

When we are able to reject $H_0$, we conclude that $\rho$ is not equal to zero and that therefore a correlation or association between the two variables exists.

**Assumptions of the Test**

1. The sample is a random sample from the population of interest.
2. Measurement of both variables is on an interval or ratio scale.
3. Both variables are approximately normally distributed.
4. The relationship between the two variables, if it exists, is linear.

When assumption 3 is not met, transformation (see chapter 10) of one or both variables may make them approximately normally distributed. However, great care must be taken here, since a transformation may cause the relationship to be nonlinear, and then assumption 4 would be violated. When assumptions 2 and/or 3 and/or 4 are not met, a nonparametric correlation test (discussed later in this chapter) should be used instead. When assumption 4 is not met, transformation of one or both variables might make the relationship linear. Again though, caution should be exercised here, since transformation to make the relationship linear may result in a nonnormal distribution. When assumption 1 is not met, correlation analysis is inappropriate. The data can't be "fixed," so a new sample must be collected.

We will use example 12.1, discussed in the previous chapter (section 12.1), to illustrate the calculations involved in correlation analysis. This example concerns the relationship between the postpartum weight of female iguanas and the number of eggs they produce in a single brood.

These data are reproduced in table 13.1 with some preliminary calculations and are shown graphically in the last chapter (figure 12.1).

**Table 13.1** Postpartum weight of female iguanas and number of eggs produced

| Specimen Number | Mass (in Kilograms) | Number of Eggs |
|---|---|---|
| 1 | 0.90 | 33 |
| 2 | 1.55 | 50 |
| 3 | 1.30 | 46 |
| 4 | 1.00 | 33 |
| 5 | 1.55 | 53 |
| 6 | 1.80 | 57 |
| 7 | 1.50 | 44 |
| 8 | 1.05 | 31 |
| 9 | 1.70 | 60 |
| 10 | 1.20 | 40 |
| 11 | 1.45 | 50 |
| | $\sum x = 15$ | $\sum y = 497$ |
| | $\sum x^2 = 21.33$ | $\sum y^2 = 23{,}449$ |
| | $(\sum x)^2 = 225$ | $(\sum y)^2 = 247{,}009$ |
| | $\sum xy = 705.8$ | |
| | $\sum x \sum y = 7{,}455$ | |

In this example, the data are measured on a ratio scale, and both variables may be assumed to be approximately normally distributed. Accordingly, we may compute the Pearson correlation coefficient ($r$) by

$$r = \frac{\sum xy - \dfrac{\sum x \sum y}{n}}{\sqrt{\left(\sum x^2 - \dfrac{(\sum x)^2}{n}\right)\left(\sum y^2 - \dfrac{(\sum y)^2}{n}\right)}} \quad (13.1)$$

The numerator in equation 13.1 is called the covariance, and it measures how $x$ and $y$ vary together. You will recognize the denominator as the square root of the product of the sums of squares of the $x$ and $y$ variables. Substituting the values from the example in equation 13.1 gives

$$r = \frac{705.8 - \left[\dfrac{(15)(497)}{11}\right]}{\sqrt{\left(21.33 - \dfrac{(15)^2}{11}\right)\left(23{,}449 - \dfrac{(497)^2}{11}\right)}} = 0.952$$

Recall that the correlation coefficient can range from –1 to +1. The value 0.952 is close to 1.00, suggesting a strong correlation between the two variables. A value of +1.00 would be a perfect correlation, with all the points falling exactly on a straight line and one variable increasing as the other variable increases. Referring back to figure 12.1 confirms that this is indeed the case. Nevertheless, we should always compare the sample statistic with critical values to assess the original null hypothesis.

### 13.1.1 Testing the Significance of *r*

Recall that the null hypothesis in a correlation analysis is that $\rho$, the population correlation coefficient, is zero. The sample correlation coefficient, $r$, is an estimate of this population correlation coefficient and has a $t$ distribution with $n - 2$ degrees of freedom. We calculate $t$ by

$$t = r\sqrt{\frac{n-2}{1-r^2}} \quad (13.2)$$

For the example,

$$t = 0.952\sqrt{\frac{9}{1-0.906}} = 9.315$$

Referring to table A.2, the critical value of $t$ for alpha = 0.05 and 9 df is 2.262 for a two-tailed probability. Since our $t$ statistic exceeds even the largest critical value at 9 df, the $p$ value is very small ($p < 0.0001$). We may therefore reject the null hypothesis and conclude that the two variables are correlated.

An alternative way of determining if $r$ is significant is to consult table A.8. This table gives the minimum values of $r$ that permit one to reject the null hypothesis. If the calculated value of $r$ is equal to or greater than the critical value for the specified degrees of freedom ($n - 2$), the null hypothesis is rejected. In the current example, 0.952 >> 0.602, leading to the same conclusion as above.

## 13.2 A Correlation Matrix

Quite often we have multiple measurement variables and would like to quickly assess which are associated with which. We might have measures of a number of environmental features and would like to know how each feature tends to associate with all other features. If we had 10 such features, there are 45 possible pair-wise combinations. This leads to a lot of number crunching. Fortunately, computers are good at this.

Statistical software can calculate correlations quite easily and typically show the results in the form of a matrix, with the statistic ($r$) and $p$ value illustrated for each pair. Such a result is shown below in table 13.2 for student scores from a recent biometry class ($n$ = 55). The five variables are listed at the top, together with their descriptive statistics. The correlation matrix is shown at the bottom for all possible pairs of these variables. (Note that one of the variables is derived from summing three of the others.) In this matrix, at each cell (an intersection of row and column position) the top number is the correlation coefficient and bottom number is the $p$ value. Since the $p$ values are all 0.000 (actually $p < 0.001$), every variable is strongly correlated to every other one. A scatterplot for two of these variables is shown in figure 13.2. Notice that, although the positive relationship between these two variables is evident, substantial unexplained variation remains.

One point of caution about correlation matrices is that the risk of a type I error increases with multiple tests (recall section 10.3.2). The simplest solution is to divide alpha (usually 0.05) by the number of tests. In this case, there are 10 pair-wise combinations of 5 variables, so we would reject $H_0$ ($\rho = 0$) only when the $p$ value is less then $0.05 \div 10 = 0.005$. That is clearly the case for all of our comparisons (table 13.2), so our earlier conclusions still hold.

## 13.3 Nonparametric Correlation Analysis (Spearman's r)

When data are measured on an ordinal scale or when other assumptions of the parametric correlation test just discussed are not met, one may use a nonparametric correlation test. The most commonly used is the Spearman rank correlation test. The assumptions of this test are that observations are a random sample from the population and that measurement of both variables is at least ordinal.

---

### Example 13.2

A nonparametric correlation

The mass (in grams) of 13 adult male tuatara and the size of their territories (in square meters) were measured. Are territory size and the size of the male holding the territory related? (In other words, do larger males hold larger territories?)

---

**Table 13.2** Descriptive statistics and correlation matrix for scores by 55 biometry students, as displayed by MINITAB. A $p$ value reported as 0.000 should be interpreted as $p < 0.001$. For $n = 55$, critical values are $r_{53,\,.05} = 0.264$, $r_{53,\,.01} = 0.365$

**Descriptive Statistics: Exam1, Exam2, Exam3, ProbSetTotal, ExamsTotal**

| Variable | N | N* | Mean | SE Mean | StDev | Minimum | Q1 | Median | Q3 |
|---|---|---|---|---|---|---|---|---|---|
| Exam1 | 55 | 1 | 81.55 | 1.41 | 10.42 | 59.00 | 74.00 | 86.00 | 89.00 |
| Exam2 | 55 | 1 | 81.45 | 1.56 | 11.58 | 58.00 | 74.00 | 83.00 | 91.00 |
| Exam3 | 54 | 2 | 79.61 | 1.49 | 10.95 | 47.00 | 73.75 | 82.00 | 88.00 |
| ProbSetTotal | 55 | 1 | 103.05 | 2.23 | 16.57 | 53.00 | 94.00 | 109.00 | 115.00 |
| ExamsTotal | 54 | 2 | 243.33 | 3.85 | 28.32 | 172.00 | 223.50 | 245.00 | 269.25 |

| Variable | Maximum |
|---|---|
| Exam1 | 95.00 |
| Exam2 | 100.00 |
| Exam3 | 98.00 |
| ProbSetTotal | 120.00 |
| ExamsTotal | 282.00 |

**Correlations: Exam1, Exam2, Exam3, ProbSetTotal, ExamsTotal**

| | Exam1 | Exam2 | Exam3 | ProbSetTotal |
|---|---|---|---|---|
| Exam2 | 0.654<br>0.000 | | | |
| Exam3 | 0.597<br>0.000 | 0.685<br>0.000 | | |
| ProbSetTotal | 0.494<br>0.000 | 0.521<br>0.000 | 0.493<br>0.000 | |
| ExamsTotal | 0.844<br>0.000 | 0.893<br>0.000 | 0.876<br>0.000 | 0.541<br>0.000 |

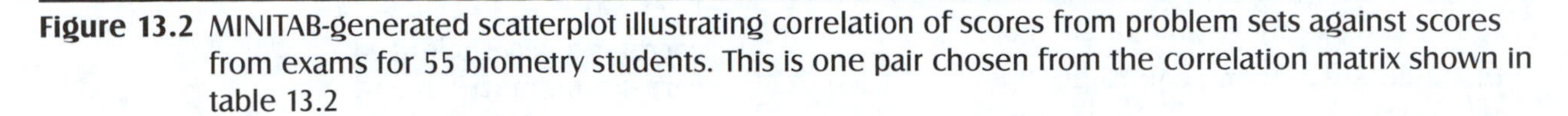

**Figure 13.2** MINITAB-generated scatterplot illustrating correlation of scores from problem sets against scores from exams for 55 biometry students. This is one pair chosen from the correlation matrix shown in table 13.2

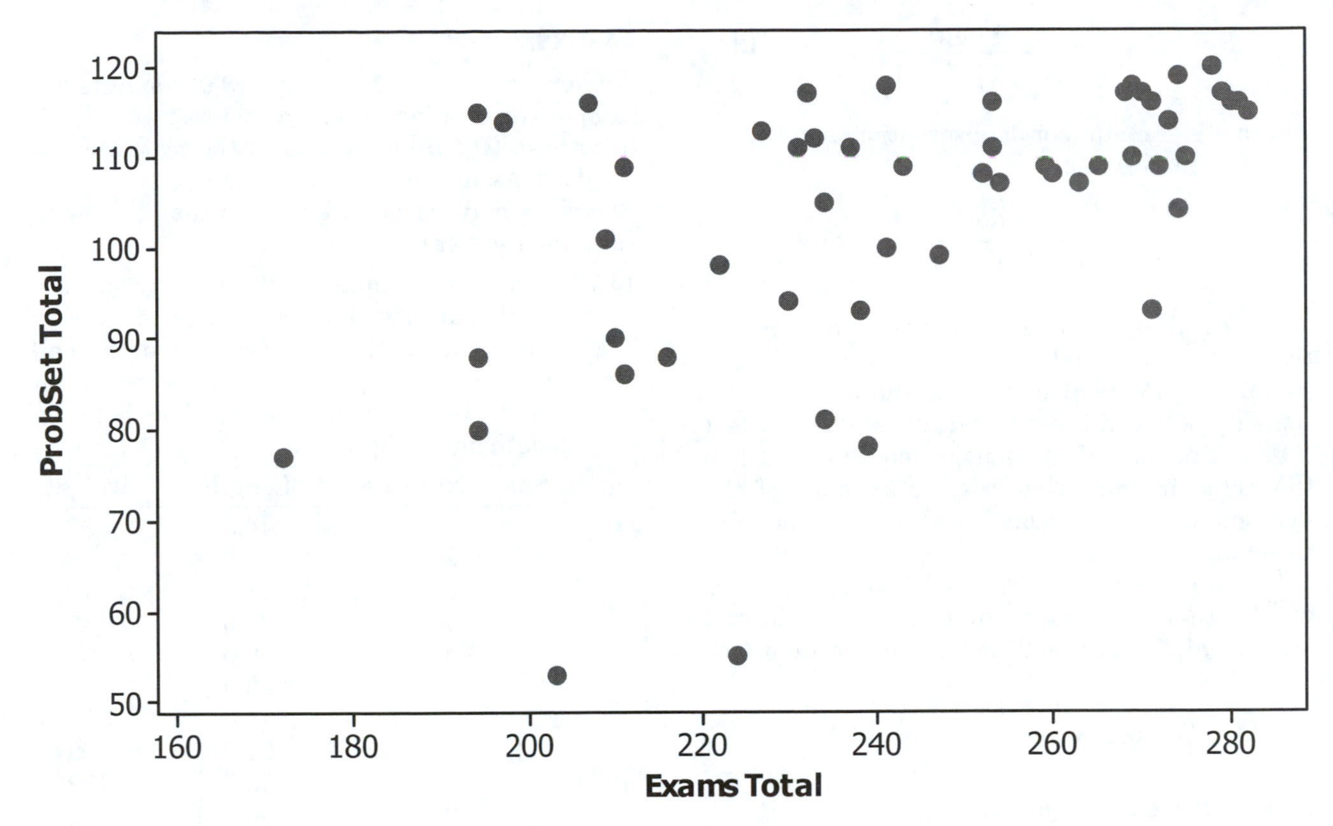


The mass of these animals is probably approximately normally distributed, but it seems doubtful that the area of their territories is, and even if it were, territory size is rather difficult to measure accurately. For these reasons a nonparametric correlation test seems to be in order. The results are shown in table 13.3. Here we arbitrarily designate mass as the $x$ variable and territory size as the $y$ variable. A scatterplot (not shown) suggests a positive correspondence between these variables.

The steps involved in calculating the Spearman correlation coefficient, $r_s$, follow. We use the tuatara data (table 13.3) as an example.

1. Rank the observations in the $x$ variable from smallest to largest. These are designated as R$x$ in the table.
2. Rank the observations in the $y$ variable from smallest to largest. These are designated as R$y$ in the table. (Note: The $x$ and $y$ variables are ranked separately.)
3. Subtract each R$y$ from its corresponding R$x$ (designated as $d$ in the table), and square each difference (designated as $d^2$).

**Table 13.3** Mass and territory size of adult male tuataras

| Observation Number | Mass ($x$) | R$x$ | Territory Size ($y$) | R$y$ | $d$ | $d^2$ |
|---|---|---|---|---|---|---|
| 1 | 510 | 6 | 6.9 | 6 | 0 | 0 |
| 2 | 773 | 9 | 20.6 | 11 | −2 | 4 |
| 3 | 840 | 13 | 17.2 | 9 | 4 | 16 |
| 4 | 505 | 5 | 6.7 | 5 | 0 | 0 |
| 5 | 765 | 8 | 20.0 | 12 | −4 | 16 |
| 6 | 780 | 10 | 24.1 | 13 | −3 | 9 |
| 7 | 235 | 1 | 1.5 | 1 | 0 | 0 |
| 8 | 790 | 11 | 13.8 | 8 | 3 | 9 |
| 9 | 440 | 3 | 1.7 | 2 | 1 | 1 |
| 10 | 435 | 2 | 2.1 | 3 | −1 | 1 |
| 11 | 815 | 12 | 20.2 | 10 | 2 | 4 |
| 12 | 460 | 4 | 3.0 | 4 | 0 | 0 |
| 13 | 697 | 7 | 10.3 | 7 | 0 | 0 |
| | | | | | | $\sum d^2 = 60$ |

Data from J. Gillingham

4. Sum the values of $d^2$, designated as $\Sigma d^2$. For the example, this value is 60.
5. Calculate $r_s$ using equation 13.3.

$$r_s = 1 - \left( \frac{6 \times \Sigma d^2}{n \times (n^2 - 1)} \right) \quad (13.3)$$

where $n$ is the number of observations.

For the example,

$$r_s = 1 - \left( \frac{6 \times 60}{13 \times (169 - 1)} \right) = 0.835$$

The null hypothesis in the Spearman correlation test is essentially the same as it is for the parametric Pearson test—that there is no relationship between the two variables. We may test the significance of $r_s$ by using equation 13.2 in the same manner that the significance of the Pearson correlation coefficient was tested. For the example,

If our computed value of $t$ is equal to or greater than the critical value of $t$ for the desired level of alpha at $n - 2$ degrees of freedom (see

$$t = 0.835 \sqrt{\frac{13 - 2}{1 - (0.835)^2}} = 4.89$$

table A.2), we may reject $H_0$.

The null hypothesis may also be tested by consulting table A.9. This table gives minimum values of $r_s$, which are significant at various degrees of freedom. When the calculated value of $r_s$ is equal to or greater than the table value, the null hypothesis is rejected. Using either approach, you should see that the null hypothesis is rejected. Using table A.2 (at 11 df), we see that $p < 0.001$. Clearly, there is a tendency for larger tuatara males to hold larger territories. From the experience of behavioral ecologists, such a result is consistent with studies from a wide variety of other animals.

## Key Terms

correlation coefficient
correlation matrix

## Exercises

For exercises 13.1 through 13.9, compute the sample correlation coefficient, test the null hypothesis ($H_0$: $\rho = 0$), and draw the scatterplot of the data. Assume a bivariate normal distribution. Exercises may be done either by hand or with statistical software.

13.1 A random sample of 28 sexually mature female garter snakes was collected. Among other things, the snout-vent length (cm) and the weight (g) of each snake were measured. Is there an association between length and weight?

| Snake Number | Length | Weight |
|---|---|---|
| 1 | 44.5 | 96 |
| 2 | 49.0 | 68 |
| 3 | 41.0 | 72 |
| 4 | 30.0 | 17 |
| 5 | 48.0 | 94 |
| 6 | 43.0 | 52 |
| 7 | 41.0 | 74 |
| 8 | 53.0 | 92 |
| 9 | 52.0 | 92 |
| 10 | 50.0 | 64 |
| 11 | 49.5 | 80 |
| 12 | 49.0 | 86 |
| 13 | 48.5 | 52 |
| 14 | 44.0 | 71 |
| 15 | 50.5 | 87 |
| 16 | 48.5 | 85 |
| 17 | 50.5 | 69 |
| 18 | 68.5 | 214 |
| 19 | 52.0 | 152 |
| 20 | 42.0 | 83 |
| 21 | 42.0 | 90 |
| 22 | 46.0 | 59 |
| 23 | 52.0 | 92 |
| 24 | 70.5 | 170 |
| 25 | 58.5 | 120 |
| 26 | 50.5 | 94 |
| 27 | 40.5 | 53 |
| 28 | 48.0 | 51 |

13.2 A random sample of female mosquito fish was collected. Their total length (mm) and the number of embryos that each contained were determined. Is there an association between length and number of embryos?

| Length | No. of Embryos | Length | No. of Embryos |
|---|---|---|---|
| 30 | 19 | 32 | 35 |
| 28 | 8 | 33 | 17 |
| 53 | 59 | 34 | 39 |
| 39 | 56 | 30 | 25 |
| 37 | 23 | 34 | 38 |
| 37 | 27 | 33 | 19 |
| 37 | 26 | 32 | 17 |
| 37 | 52 | 30 | 17 |
| 39 | 59 | 32 | 26 |
| 36 | 31 | 32 | 16 |
| 33 | 19 | 29 | 18 |
| 35 | 50 | 30 | 22 |
| 38 | 53 | 31 | 21 |
| 39 | 22 | 36 | 17 |
| 30 | 9 | 33 | 20 |

13.3 The age, systolic blood pressure, and diastolic blood pressure for a random sample of 62 people were determined. Is there a correlation between age and systolic blood pressure? Age and diastolic blood pressure? Diastolic and systolic blood pressure?

| Age | Systolic | Diastolic | Age | Systolic | Diastolic |
|---|---|---|---|---|---|
| 22 | 114 | 74 | 21 | 112 | 68 |
| 20 | 118 | 68 | 25 | 111 | 70 |
| 7 | 94 | 54 | 7 | 90 | 53 |
| 10 | 94 | 48 | 9 | 92 | 56 |
| 30 | 118 | 64 | 21 | 120 | 70 |
| 21 | 140 | 70 | 24 | 120 | 80 |
| 35 | 118 | 78 | 37 | 138 | 84 |
| 38 | 120 | 80 | 29 | 123 | 78 |
| 10 | 100 | 40 | 41 | 132 | 83 |
| 22 | 125 | 90 | 22 | 110 | 77 |
| 15 | 108 | 58 | 11 | 102 | 66 |
| 46 | 130 | 90 | 19 | 108 | 74 |
| 39 | 130 | 94 | 79 | 140 | 86 |
| 22 | 104 | 62 | 22 | 132 | 72 |
| 58 | 134 | 72 | 24 | 138 | 82 |
| 46 | 122 | 86 | 37 | 122 | 78 |
| 37 | 142 | 82 | 26 | 132 | 64 |
| 40 | 122 | 62 | 40 | 125 | 63 |
| 37 | 110 | 72 | 9 | 90 | 42 |
| 27 | 110 | 64 | 20 | 122 | 78 |
| 21 | 110 | 68 | 21 | 106 | 62 |
| 22 | 138 | 82 | 51 | 118 | 78 |
| 20 | 118 | 62 | 20 | 120 | 78 |
| 19 | 118 | 76 | 36 | 132 | 76 |

*(continued)*

| Age | Systolic | Diastolic | Age | Systolic | Diastolic |
|---|---|---|---|---|---|
| 53 | 114 | 78 | 18 | 118 | 80 |
| 43 | 112 | 78 | 21 | 115 | 70 |
| 33 | 116 | 79 | 22 | 114 | 70 |
| 36 | 117 | 78 | 13 | 106 | 66 |
| 21 | 114 | 76 | 25 | 114 | 66 |
| 41 | 116 | 74 | 21 | 110 | 70 |
| 38 | 102 | 62 | 7 | 80 | 42 |

13.4 It is likely that students who do well on an examination in a particular course are likely to score well on a subsequent examination in that same course, and that students who do not do well on the first exam are likely to not do well on the second exam (i.e., exam scores on the two exams will be correlated). Below are the scores on two examinations for 45 randomly selected students in a general biology course. Is there an association between the scores on the two tests?

| Test 1 | Test 2 | Test 1 | Test 2 | Test 1 | Test 2 |
|---|---|---|---|---|---|
| 63 | 76 | 49 | 71 | 66 | 67 |
| 51 | 48 | 46 | 69 | 72 | 57 |
| 46 | 64 | 54 | 62 | 74 | 74 |
| 74 | 88 | 74 | 88 | 74 | 83 |
| 83 | 88 | 40 | 64 | 83 | 83 |
| 80 | 86 | 74 | 83 | 83 | 86 |
| 49 | 79 | 72 | 60 | 83 | 83 |
| 89 | 79 | 72 | 62 | 80 | 57 |
| 77 | 74 | 51 | 52 | 80 | 62 |
| 86 | 79 | 66 | 74 | 72 | 50 |
| 46 | 79 | 89 | 88 | 83 | 69 |
| 60 | 74 | 60 | 79 | 77 | 60 |
| 66 | 83 | 74 | 64 | 92 | 88 |
| 66 | 71 | 72 | 74 | 86 | 74 |
| 54 | 81 | 69 | 76 | 60 | 62 |

13.5 10 randomly selected soil samples were analyzed for krypton content ($\mu$g/kg soil) using an old, expensive, but very reliable and accurate method and a newer, less expensive, faster method. A strong correlation between the results of the two methods would indicate that the new method is also accurate.

| Old Method | New Method |
|---|---|
| 25 | 27 |
| 30 | 28 |
| 20 | 19 |
| 35 | 36 |
| 40 | 38 |
| 25 | 25 |
| 33 | 32 |
| 50 | 52 |
| 65 | 67 |
| 60 | 58 |

13.6 Total body weight (g), spleen weight (mg), and bursa weight (mg) of 9 newly hatched turkeys were determined. Is there a correlation between total body weight and spleen weight? Total body weight and bursa weight? Spleen weight and bursa weight?

| Body Weight | Spleen Weight | Bursa Weight |
|---|---|---|
| 53.81 | 18.9 | 50.8 |
| 56.26 | 20.4 | 51.4 |
| 59.86 | 15.9 | 28.4 |
| 59.96 | 19.9 | 66.6 |
| 61.75 | 17.4 | 35.5 |
| 55.28 | 24.0 | 38.8 |
| 56.57 | 21.3 | 50.3 |
| 49.91 | 16.2 | 33.2 |
| 54.25 | 19.3 | 39.3 |

13.7 The weight, length, and width of 44 randomly selected killdeer eggs were determined. Is there a correlation between weight and width? Width and length? Weight and length? (Data from D. Blaszkiewicz.)

| Weight (g) | Width (mm) | Length (mm) | Weight (g) | Width (mm) | Length (mm) |
|---|---|---|---|---|---|
| 13 | 26.4 | 36.7 | 13.5 | 27.4 | 37.4 |
| 13 | 26.5 | 36.5 | 13.5 | 26.9 | 38.5 |
| 12 | 26.4 | 34.3 | 14 | 27.4 | 37.8 |
| 13.5 | 27.1 | 37.1 | 15.5 | 28.7 | 38 |
| 15.5 | 28.3 | 38.1 | 15.5 | 28.4 | 39 |
| 15 | 28 | 37.2 | 16 | 28.4 | 39 |
| 15 | 28 | 38.1 | 12.5 | 26.3 | 37.6 |
| 14 | 27.2 | 37.5 | 13 | 26.5 | 37.6 |
| 14 | 27.7 | 36.7 | 12 | 25.9 | 37.1 |
| 15 | 27.6 | 38.3 | 14 | 27.1 | 38.3 |
| 14.5 | 28 | 36.5 | 15 | 27 | 42 |
| 14.5 | 27.2 | 39.5 | 14.5 | 27.3 | 39.6 |
| 14.5 | 26.8 | 39.1 | 14 | 27 | 37.8 |
| 13 | 26 | 39.6 | 15 | 27.2 | 40.2 |
| 13 | 26 | 36.9 | 12 | 27.5 | 35.3 |
| 12.5 | 27.3 | 36.4 | 12.5 | 27.3 | 34.3 |
| 12 | 26.7 | 36 | 13 | 26.3 | 38.1 |
| 12.5 | 27.3 | 36.5 | 13 | 26.4 | 38.2 |
| 12.5 | 26.8 | 37.1 | 13 | 28.3 | 39.3 |
| 14.5 | 27.2 | 37.6 | 14 | 29.4 | 38.7 |
| 14.5 | 27.4 | 39.5 | 15 | 28.4 | 39.6 |
| 13.5 | 27.8 | 38.1 | 14 | 27.7 | 39.3 |

13.8 Using the data in digital appendix 3, determine if there is a correlation between pulse rate and reaction time in humans.

13.9 A behavioral ecologist tested the hypothesis that larval ringed salamanders (*Ambystoma annulatum*) respond to chemical cues from predatory newts (*Notophthalmus viridescens*) by reducing their activity levels and that those most at risk to predation show the largest response. Prior studies indicated that newts have more difficulty eating larger salamanders than smaller salamanders; thus we would expect that, the smaller the salamander, the greater the risk of mortality to these predators. The ecologist measured the "change in activity" as the time active after exposure to the cue minus the time active before exposure. Positive numbers indicate increased activity, whereas negative numbers indicate decreased activity. She then conducted an experiment to test the hypothesis that smaller salamanders show the largest decrease in activity. Based on prior studies, we know that such activity data show a strong departure from normality. Use the appropriate test on the following data ($n$ = 16) to see if the correlation is statistically discernable. If so, is the correlation in the direction expected? (Data from A. Mathis.)

| Size (mm) | Change in activity | Size (mm) | Change in activity |
|---|---|---|---|
| 2.4 | 65 | 2.3 | –148 |
| 1.5 | –7 | 2.2 | –12 |
| 2.7 | 99 | 2.5 | –69 |
| 2.5 | 75 | 2.6 | –32 |
| 2.4 | –55 | 2.0 | 18 |
| 2.5 | 121 | 2.8 | 68 |
| 2.0 | –47 | 2.0 | –188 |
| 1.9 | –150 | 2.1 | –248 |

# 14

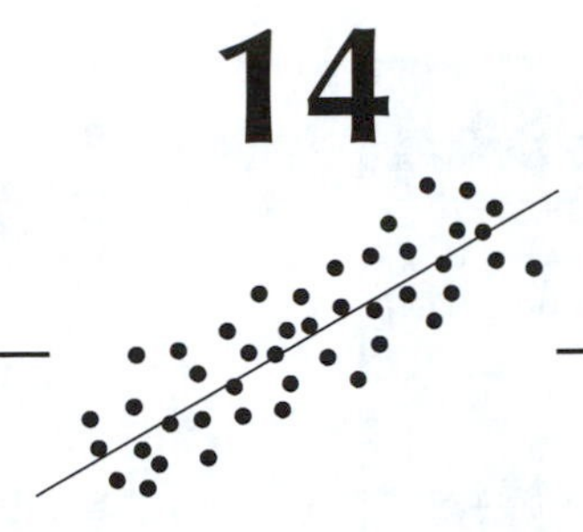

# Analysis of Frequencies

In biological research, we encounter data from different kinds of variables and scales of measurement (sections 2.1 and 2.2). So far in this text we have emphasized parametric tests on continuous data measured on an interval or ratio scale or their nonparametric counterparts, which may also be applied to data from an ordinal scale. However, in much biological and medical research we are interested in the frequencies with which events or objects occur, as they are classified into different attributes (nominal scale). In genetics we commonly ask whether the frequencies of observed phenotypes conform to an expected distribution from a particular mode of inheritance. In medicine, epidemiologists and other public-health workers frequently ask about the distribution of diseases in populations. For example, does the incidence of hepatitis-A differ between people in Chicago and people in St. Louis? In this chapter, we consider statistical tests used to analyze frequency data. We are generally interested in two types of problems: checking for goodness-of-fit and checking for differences between groups. We use the most common procedure, called the chi-square test, for both types of problems. We also introduce two other procedures for other situations.

## 14.1 The Chi-Square Goodness-of-Fit Test

We often want to determine if a frequency distribution of a sample fits or does not fit some expected theoretical distribution, such as a Poisson, binomial, or normal distribution or, for that matter, any sort of distribution we care to specify. Several tests, called **goodness-of-fit tests**, are designed for just this purpose.

The null hypothesis for a goodness-of-fit test is that an observed frequency distribution is not different from some specified distribution and that any departure of the frequency distribution of the sample from the specified distribution is therefore due to chance alone.

You are probably already acquainted with the chi-square goodness-of-fit test from genetics. The chi-square goodness-of-fit test determines how well a set of observed frequencies in two or more mutually exclusive categories fit (or do not fit) some specified expected distribution. It is a nonparametric test and requires only nominally measured data plus their frequencies. Other variable types can also be used after pooling them into categories (e.g., frequency distributions of continuous measurements).

**Assumptions of the Test**

1. Measurement is on at least a nominal scale. Categories of the nominal scale or of the groups whose frequencies are represented are mutually exclusive.
2. Observations are independent.
3. No category has an expected frequency of less than 5 (or when there are many categories, not more than 20% of the categories have an expected frequency of less than 5).

## Example 14.1

### A chi-square goodness-of-fit test

Two purple-flowered pea plants, both heterozygous for flower color, were crossed, resulting in 78 purple-flowered offspring and 22 white-flowered offspring. Does this outcome differ from a 3:1 ratio of purple-flowered to white-flowered offspring?

Color is controlled by a single pair of alleles, and purple is dominant over white. Because of this we expect a ratio of 3 purple-flowered offspring to 1 white-flowered offspring, based on a binomial probability (section 5.1). Thus, we would expect 75 purple-flowered plants and 25 white-flowered plants (which is the null hypothesis). However, by chance alone we might expect that there would be some discrepancy between the expected results and what we actually obtained.

A goodness-of-fit test helps us to determine whether this observed discrepancy between the observed frequencies and the expected frequencies too large to be attributed to chance alone. In other words, is the difference between the observed result and the expected result significant (statistically discernable)? The chi-square statistic is given by

$$\chi^2 = \Sigma \frac{(o-e)^2}{e} \tag{14.1}$$

The steps in its computation, shown in table 14.1, are as follows.

1. Subtract each expected value (column 2) from each observed value (column 1). The difference $(o - e)$ is shown in column 3.
2. Square the values of $(o - e)$ to obtain the squared differences, $(o - e)^2$, as shown in column 4.
3. Divide each squared difference by the expected value $(e)$ for that category to obtain the value of $(o - e)^2/e$ for each category (column 5).
4. Sum the values obtained in column 5 to obtain the chi-square statistic.
5. The degrees of freedom for a chi-square goodness-of-fit test is the number of mutually exclusive groups minus one.

**Table 14.1** Calculation of the chi-square statistic

| 1<br>observed | 2<br>expected | 3<br>$(o-e)$ | 4<br>$(o-e)^2$ | 5<br>$\frac{(o-e)^2}{e}$ |
|---|---|---|---|---|
| 78 | 75 | 3 | 9 | 0.12 |
| 22 | 25 | −3 | 9 | 0.36 |

$$\chi^2 = \Sigma \frac{(o-e)^2}{e} = 0.48$$

We now consult table A.3. The top row gives probabilities associated with values of chi-square, and the left column gives degrees of freedom. Our calculated value, 0.48 with one degree of freedom, is less than the critical value for alpha = 0.05 and 1 degree of freedom. Therefore, we are unable to reject the null hypothesis. We conclude that the observed frequency does not differ from the expected frequency, which confirms our genetic model. Now let us look at a somewhat more complex situation.

## • Example 14.2

### A chi-square goodness-of-fit test

Refer to the maple-seedlings-per-quadrat example from section 5.2 (Poisson distribution), which is reproduced here as table 14.2. We wish to know if the seedlings are randomly or non-randomly distributed in the sampled habitat.

If the maple seedlings in the sampled area were randomly distributed, we would expect, on the basis of the Poisson distribution, to obtain the results given in the column headed "Expected." These values are derived from the null hypothesis that the plant locations will follow a Poisson distribution. If we reject the null hypothesis, we accept the alternative hypothesis that the seedlings are not randomly distributed. The observed results are shown in the column headed "Observed." We compute the chi-square as before, with one small difference. For the chi-square test

**Table 14.2** Expected and observed values of maple seedlings per quadrat ($n = 100$, $\bar{x} = 1.41$). Note that the last three rows are pooled for calculating chi-square

| Seedlings/ Quadrat | Observed | | Expected | | $\frac{(o-e)^2}{e}$ |
|---|---|---|---|---|---|
| 0 | 35 | | 24.41 | | 4.59 |
| 1 | 28 | | 34.42 | | 1.20 |
| 2 | 15 | | 24.27 | | 3.54 |
| 3 | 10 | | 11.41 | | 0.17 |
| 4 | 7 | | 4.02 | | — |
| 5 | 5 | 12 | 1.13 | 5.48 | 7.76 |
| ≥ 6 | 0 | | 0.33 | | — |

$$\chi^2 = \sum \frac{(o-e)^2}{e} = 17.26$$

Data from table 5.3

to be valid, not more than 20% of the expected values should be less than 5 (assumption 3), and some of our expected values fall below this. We solve this problem simply by combining the last three values in both the observed and expected columns (pooling the data). Table 14.2 gives the results of the chi-square computations for this example. Check to see that the last column and $\chi^2$ are calculated correctly.

There appear to be 4 degrees of freedom in this case (5 groups – 1). However, when it is necessary to estimate one or more parameters in the population of interest, one degree of freedom must be deducted for each parameter estimated. Recall that in this case our expected values are based on our sample mean of 1.41 seedlings per quadrat. In other words, we use an estimate of the population mean to obtain the expected values. Thus, the degrees of freedom are 3 rather than 4. Reference to table A.3 gives a critical value of $\chi^2$ at alpha = 0.05 and 3 degrees of freedom as 7.82. Since our calculated value is larger than this, we reject the null hypothesis. The $p$ value is $0.0001 < p < 0.001$. We thus conclude that the observed distribution does not follow a Poisson distribution and hence also conclude that the maple seedlings are not randomly distributed in the sampled habitat.

## 14.2 The Chi-Square Test for Association

**Tests for association** are used to determine if two variables, both measured on a nominal scale, are related or associated in some way. (These are also called **heterogeneity chi-square** tests and chi-square tests of independence.) It is helpful to think of these tests as tests for "correlation" between nominal variables. An example might help clarify this.

### • Example 14.3

A chi-square test for association

In certain parts of west Africa where malaria is prevalent, there is a mutant form of hemoglobin called sickle-cell hemoglobin or hemoglobin S. Individuals who are homozygous for the hemoglobin-S allele develop sickle-cell anemia, an often fatal disease. Heterozygous individuals exhibit some mild symptoms of anemia but have an abnormally high resistance to the malaria parasite. Individuals who are homozygous for normal hemoglobin are highly susceptible to malaria. The now infamous experiment that confirmed this demonstrates well the relationship between two nominal variables—genotype and susceptibility to malaria—although the experiment itself would be regarded as inhumane and unethical by present standards. In this experiment 30 prison inmates, all volunteers, were selected. Fifteen of these individuals were heterozygous for hemoglobin S, and 15 were homozygous for normal hemoglobin. All 30 were from the same general population and were therefore genetically similar in other respects. All 30 were artificially infected with the malaria parasite. Of the 15 homozygous individuals, 13 contracted the disease, and two did not. Of the 15 heterozygous individuals, one contracted the disease, and the remaining 14 did not. Such data are customarily arrayed in the form of a matrix, sometimes called a **contingency table**, as shown in table 14.3.

**Table 14.3** Association between the hemoglobin-s allele and resistance to malaria

| | **Contracted malaria** | **Did not contract malaria** | **Totals** |
|---|---|---|---|
| Heterozygotes | 1 | 14 | 15 |
| Homozygotes | 13 | 2 | 15 |
| Totals | 14 | 16 | 30 |

Such an experiment allows us to determine whether there is a relationship between the two variables. Remember, the two variables in this case are genotype (measured on a nominal scale) and susceptibility to malaria (also measured on a nominal scale). The null hypothesis is that there is no relationship (i.e., that the two variables are independent).

Consider the experiment just outlined in example 14.3. The results are given again in table 14.4, with the expected values in parentheses.

**Assumptions of the Test**

1. Data are frequencies.
2. Samples are independent (i.e., the same individual may not occur in more than one cell).
3. Not more than 20% of the cells may have an expected value of <5, and no cell may have an expected value <1. For a 2 × 2 contingency table, all cells must have an expected value of 5 or greater.

**Table 14.4** Association between the hemoglobin-s allele and resistance to malaria

| | Contracted malaria | Did not contract malaria | Totals |
|---|---|---|---|
| Heterozygotes | 1 (7.00) | 14 (8.00) | 15 |
| Homozygotes | 13 (7.00) | 2 (8.00) | 15 |
| Totals | 14 | 16 | 30 |

The expected value for each cell in a test for association is obtained by multiplying the row total by the column total for a cell, and then dividing the product by the grand total, or

$$\frac{\text{Row Total} \times \text{Column Total}}{\text{Grand Total}} \tag{14.2}$$

For instance, the expected frequency of heterozygotes contracting malaria is:

$$\frac{15 \times 14}{30} = 7.00$$

By the way, this formula for computing expected frequencies follows the multiplication rule of probability $P(A \times B) = P(A) \times P(B)$ (section 5.1), which is true when the component probabilities are independent!

We next compare the observed result with the result predicted by the null hypothesis, which is that there is no relationship between the heterozygous condition for this gene and resistance to malaria. We calculate a value for chi-square as before: by summing the values $(o - e)^2/e$ for each of the cells. Chi-square for this example is

$$\chi^2 = \frac{(1-7)^2}{7} + \frac{(14-8)^2}{8} + \frac{(13-7)^2}{7} + \frac{(2-8)^2}{8} = 19.286$$

The degrees of freedom for a chi-square test for association are:

$$\text{df} = (\text{rows} - 1) \times (\text{columns} - 1) \tag{14.3}$$

In this case $(2 - 1) \times (2 - 1) = 1$. Consulting table A.3, we find that the critical value of chi-square at alpha = 0.05 and 1 degree of freedom is 3.84. Since 19.286 > 3.84, we reject the null hypothesis. Placing our statistic on the table at 1 df, we determine the $p$ value is $p < 0.0001$. We thus conclude that there is a distinctive association between heterozygosity for the sickle-cell gene and resistance to malaria. In our initial question, we wanted to determine if the heterozygous genotype conferred some resistance to malaria. Using the chi-square test of association, we determined simply that genotype is associated with resistance. We cannot say the direction of the association, although, looking at the table, the data conform to our initial expectation (heterozygosity confers resistance). Most scientists stop here, simply showing the contingency table, the chi-square statistic, and the $p$ value and reporting the direction of change.

The chi-square test for association is not limited to 2 × 2 contingency tables, but may be expanded to as many rows and columns as necessary. Each row or each column represents one mutually exclusive category into which an individual might fall. The procedure is exactly the same when there are more than two mutually exclusive categories, except that the degrees of freedom are greater than one. Example 14.4 illustrates this.

## • Example 14.4

A chi-square test for association

An animal behaviorist wishes to determine if the performance of three different olfactory behaviors, which bulls of a certain breed of cattle typically exhibit on approaching a female of their species, is related to the reproductive stage of the female. We designate these behaviors as A, B, and C, and the reproductive stages of the females as conceptive or nonconceptive. We need not be concerned here with what these three behaviors are—only that they are mutually exclusive events. The data are given in table 14.5, with expected values in parentheses.

**Table 14.5** Frequencies of three olfactory behaviors directed to cows by bulls

| | Behavior of Bull | | | |
|---|---|---|---|---|
| **Female stage** | **A** | **B** | **C** | **Total** |
| Conceptive | 29 (29.78) | 48 (56.64) | 27 (17.58) | 104 |
| Nonconceptive | 32 (31.22) | 68 (59.36) | 9 (18.42) | 109 |
| Total | 61 | 116 | 36 | 213 |

Expected values, obtained by multiplying row totals by column totals and dividing by the grand total, are shown in parentheses. Chi-square is computed as before. In this case,

$$\chi^2 = \frac{(29-29.78)^2}{29.78} + \frac{(48-56.64)^2}{56.64} + \frac{(27-17.58)^2}{17.58} + \frac{(32-31.22)^2}{31.22} + \frac{(68-59.36)^2}{59.36} + \frac{(9-18.42)^2}{18.42} = 12.485$$

The degrees of freedom in this case are:

$$(r-1) \times (c-1) = (2-1) \times (3-1) = 2$$

As before, we consult table A.3, using 2 degrees of freedom, to determine if we may reject the null hypothesis that there is no association between the reproductive state of the female and the performance of these three behaviors by bulls. Do we accept or reject $H_0$? What is the *p* value? What can we conclude about the effects of female reproductive stage on male behavior?

Chi-square tests are very easy to do with computer programs. A goodness-of-fit test can be simply done with a spreadsheet. (This is very handy when doing multiple practice exercises; just replace the raw numbers!) MINITAB does the chi-square test of association and shows all the elements of the analysis, as well as reporting the *p* value and showing each term's contribution to the chi-square statistic. The MINITAB output for the example just conducted is displayed in table 14.6.

## 14.3 The Fisher Exact Probability Test

The Fisher exact probability test, like the chi-square test for association, is used to test for an association between two variables measured on a nominal scale. An important difference is that the requirement that no cell have an expected value of less than 5 is not applicable to the Fisher test, and, accordingly, the test is very useful for small samples. The assumption of independence (i.e., that no individual may occupy more than one cell) must be satisfied.

### • Example 14.5

**The Fisher exact probability test**

An experiment was conducted to determine if neonatal garter snakes were more or less likely to exhibit an avoidance response to a threatening stimulus presented from above or from "snake's-eye level." One sample of seven snakes was presented one at a time with an overhead stimulus, and another sample of seven was presented with the same stimulus laterally. The investigator recorded whether the snakes responded by attempting to "escape" or failed to respond. (Data from R. Hampton and J. Gillingham.)

2 independent samples are involved here—snakes that were stimulated from above and snakes that were stimulated laterally—and 2 mutually exclusive categories—responded or did not respond. Measurement is nominal. To conduct the Fisher test, data are arranged in a 2 × 2 contingency table, as in tables 14.7 and 14.8. Notice that, because of the small sample sizes, the chi-square test of association is inappropriate.

**Table 14.6** Chi-square test of association results from MINITAB for the data from table 14.5

**Chi-Square Test: A, B, C**

```
Expected counts are printed below observed counts
Chi-Square contributions are printed below expected counts

            A       B       C   Total
1          29      48      27     104
        29.78   56.64   17.58
        0.021   1.318   5.051

2          32      68       9     109
        31.22   59.36   18.42
        0.020   1.257   4.819

Total      61     116      36     213

Chi-Sq = 12.485, DF = 2, P-Value = 0.002
```

**Table 14.7** Arrangement of data for the Fisher exact probability test

| | Classification I | Classification 2 | Totals |
|---|---|---|---|
| Group I | A | B | A + B |
| Group II | C | D | C + D |
| Totals | A + C | B + D | A + B + C + D |

The probability of observing this particular distribution if the null hypothesis of no association is true is given by

$$p = \frac{(A + B)\,!\,(C + D)\,!\,(A + C)\,!\,(B + D)\,!}{n!\; A!\; B!\; C!\; D!} \quad (14.4)$$

where $n = A + B + C + D$, and by convention $0! = 1$.

The data for the example are shown in table 14.8.

**Table 14.8** Response of neonatal garter snakes to overhead and lateral stimuli

| | Responded | No Response | Totals |
|---|---|---|---|
| Overhead Stimulus | 6 | 1 | 7 |
| Lateral Stimulus | 1 | 6 | 7 |
| Totals | 7 | 7 | 14 |

The probability of this outcome under the null hypothesis is

$$p = \frac{7! \times 7! \times 7! \times 7!}{14! \times 6! \times 1! \times 1! \times 6!} = 0.01428$$

This is the probability of the observed distribution if $H_0$ is true. However, we are not interested in exactly this outcome, but rather this or any more extreme outcomes with the same marginal totals. There is only one more extreme outcome in this case:

| | Responded | No Response | Totals |
|---|---|---|---|
| Overhead Stimulus | 7 | 0 | 7 |
| Lateral Stimulus | 0 | 7 | 7 |
| Totals | 7 | 7 | 14 |

The probability for this case is

$$p = \frac{7! \times 7! \times 7! \times 7!}{14! \times 7! \times 0! \times 0! \times 7!} = 0.00029$$

Thus, the probability under $H_0$ of our observed distribution or of a more extreme one is $0.01428 + 0.00029 = 0.01457$. We may reject $H_0$ and conclude that neonatal garter snakes are more responsive to the overhead stimulus than to the eye-level stimulus.

Notice that the expected values in the above example are too small for the chi-square test for association, so we used the Fisher exact probability test. The Fisher test may also be used when the expected values are large enough to permit use of the chi-square test for association. However, the cumbersome calculations of the Fisher exact test make the chi-square test a much simpler alternative!

**Caution**

An important assumption of all the tests for association we have considered in this chapter is that observations are independent. If the data do not meet this assumption, the probability of a type I error may actually be much higher than the test result indicates.

Independent, in this case, means that no individual may occur in more than one cell of a contingency table. What exactly does this mean? Consider example 14.4, regarding the frequency with which bulls exhibited one of three behaviors in the presence of conceptive and nonconceptive cows. The total number of occurrences recorded in table 14.5 was 213. Thus, we would assume that 213 bulls were observed one time, each on approaching a cow in one of the two conditions of the test. However, only 14 bulls were used in the experiment. Accordingly, each bull was observed an average of just over 15 times! This in itself is a serous violation of independence, but to make matters worse, we have no knowledge of how many times each bull was actually observed. Some may have been observed many times (meaning that they would make a large contribution to the results), while others might have been observed only a few times. This inappropriate pooling of the data causes the test to be very unreliable. (See also section 1.4 for a further discussion of independence.)

## 14.4 The McNemar Test for the Significance of Changes

In some situations it is either not possible or not desirable to maintain strict statistical independence, as we point out in conjunction with the paired $t$-test in section 9.6 and with repeated measures ANOVA in section 11.1. We might, for example, wish to expose the same set of individuals to two different treatments and measure their responses, thus reducing the effect of variation between individuals that is not related to the treatments; or we might wish to measure a before-and-after treatment effect of some sort, measuring the response of each individual before the treatment and again after the treatment, in effect using each individual as its own control. In such cases, provided that we have designed the experiment properly, it is possible to use the McNemar test. In this test we focus on the number of individuals who change in response to the two treatments.

**Assumptions of the McNemar Test**

1. Data are measured on at least a nominal scale.
2. Each subject is measured twice: once under each of the two treatments.

## • Example 14.6

### The McNemar test

Rattlesnakes kill their prey by striking and injecting venom. Sometimes they grasp the prey immediately and wait for it to die before ingesting it; at other times they release the prey after striking it, wait for it to die, and then grasp and ingest it. This decision seems to be related to the size of the prey. Presumably, a large prey item would be capable of inflicting injury to the snake while it is dying if the snake attempted to hold the prey in its mouth during this period, while a small prey item would be less likely to inflict injury. This hypothesis predicts that large prey should be struck and released, while small prey should be struck and held.

Fifteen rattlesnakes (*Crotalus atrox*) were each fed a mouse (small prey) and a rat (large prey). It was noted whether they struck and released or struck and held the prey. The results are shown in table 14.9.

**Table 14.9** Data for example 14.6

| | | Rat (Large Prey) | | | |
|---|---|---|---|---|---|
| | | Released | | Held | |
| **Mouse** | **Released** | 3 | (*a*) | 0 | (*b*) |
| **(Small Prey)** | **Held** | 10 | (*c*) | 2 | (*d*) |

Those snakes that changed their response between the two treatments (large and small prey) are in cells $b$ and $c$. Those that released small prey but held large prey occur in cell $b$ (none did this). Those that held small prey but released large prey occur in cell $c$ (10 did this). Animals in cells $a$ and $d$ either released both sizes (cell $a$) or held both sizes (cell $d$), which is to say that they did not change their response to the two treatments. In the McNemar test, we are interested only in the "changers."

The test statistic for the McNemar test is chi-square, which is computed as

$$\chi^2 = \frac{(c-b)^2}{(c+b)} \tag{14.5}$$

For the example,

$$\chi^2 = \frac{100}{10} = 10.0$$

There is $(r-1) \times (c-1) = 1$ degree of freedom. As before, if the calculated value of chi-square is greater than the critical value with one degree of freedom, we may reject the null hypothesis that there is no change in individual responses to the two treatments. What can we conclude from this test?

When frequencies are sufficiently small such that $(c + b)/2 < 5$, we must compute a binomial probability for the McNemar test. For this binomial probability $k = c + b$, $x =$ the smaller of the two frequencies in cells $c$ or $b$, and $p = 0.5$.

## • Example 14.7

Suppose that the outcome of our rattlesnake experiment described in example 14.6 had been as follows:

| | | Rat (Large Prey) | | | |
|---|---|---|---|---|---|
| | | Released | | Held | |
| **Mouse** | **Released** | 3 | (*a*) | 0 | (*b*) |
| **(Small Prey)** | **Held** | 9 | (*c*) | 3 | (*d*) |

The value for $(c + b)/2$ is now 4.5, and the use of chi-square is therefore inappropriate. We may use the binomial test by setting $k = c + b = 9$, $x = 0$ (the smaller of the two frequencies in cells $b$ and $c$), and $p = 0.5$. Substituting these values in the binomial probability equation (section 5.1) gives

$$p(x) = \frac{k!}{x!(k-x)!} p^x q^{(k-x)}$$

$$p(0) = \frac{9!}{0! \times 9!}(0.5^0)(0.5^9) = 0.00195$$

and we may reject the null hypothesis.

## 14.5 Graphic Displays of Frequency Data

Frequency data are typically displayed as bar graphs. We can see presentation of the observed and expected frequencies of maple seedlings as a bar graph in figure 5.6 (later discussed in section 14.1). Such figures are very effective for showing departures of observed from expected distributions.

Bar graphs are also effective for revealing differences in distributions between groups such as displayed in contingency tables (section 14.2 above). For example, the frequencies of bull behavior (table 14.5) could be shown as comparative bar graphs (figure 14.1). After converting the percentages, these data may also be shown as pie graphs. (figure 14.2). Both graph types are widely used in the popular press, as well as in scientific publications.

**Figure 14.1** Frequency of olfactory behaviors of bulls, classified by reproductive stage of female

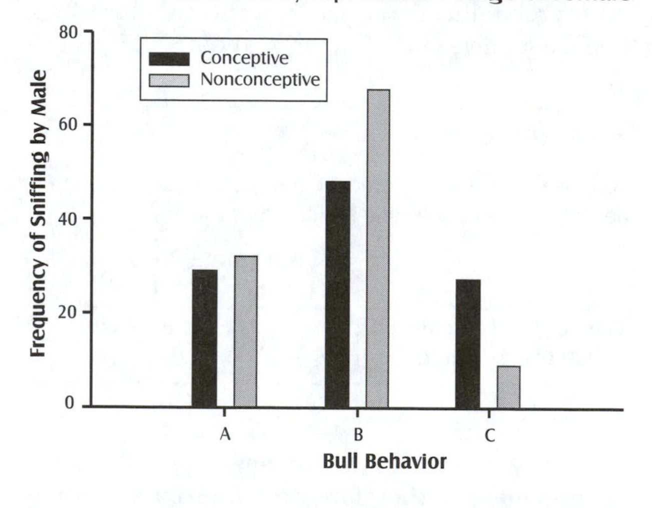


**Figure 14.2** Frequency of three different olfactory behaviors by bulls directed to cows at two different reproductive stages

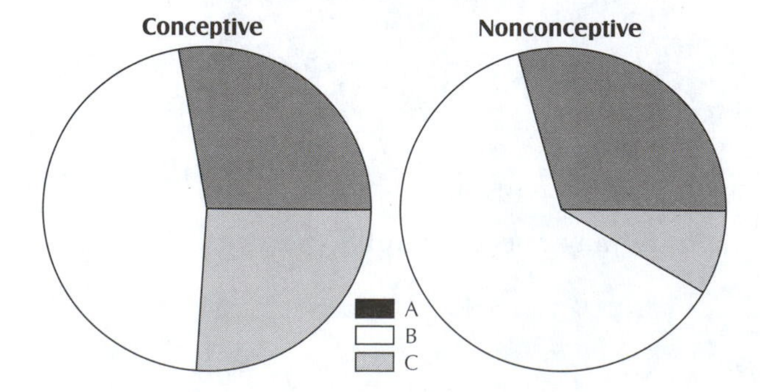


### Key Terms

chi-square test of association
contingency table
goodness-of-fit test
heterogeneity chi-square
test for association

### Exercises

Exercises preceded by an asterisk are suitable for solution using a computer.

**Chi-square Goodness-of-Fit**

14.1 100 tweetie birds were given a choice of either striped sunflower seeds or black sunflower seeds. 75 chose black seeds. May we conclude that the population from which this sample was taken has a preference for black sunflower seeds over striped sunflower seeds?

14.2 Suppose that in the situation described in exercise 14.1, 48 tweetie birds chose black seeds, and 52 chose striped seeds. May we conclude that there is a preference for striped seeds?

14.3 In a genetic experiment involving flower color in a certain plant species, a ratio of 3 blue-flowered plants to 1 white-flowered plant was expected. The observed results were 35 blue-flowered plants and 14 white-flowered plants. Does the observed ratio differ significantly from the expected ratio?

14.4 Suppose that in exercise 14.3 a ratio of 61 blue-flowered plants to 22 white-flowered plants was observed. Does the observed ratio differ significantly from the expected ratio?

14.5 We suspect that a certain strain of laboratory rats has a genetic tendency to make left turns in a "T" maze. Of 12 rats that were tested in such a maze, 8 chose to go into the left arm and 4 chose the right arm. Does this result support our suspicion about a left-turning tendency?

14.6 In 102 tosses of 4 coins, the following results were obtained. [Note that you'll need to use the binomial distribution (section 5.1) to determine expected frequencies.]

| Heads | Tails | Observed Frequency |
|---|---|---|
| 4 | 0 | 8 |
| 3 | 1 | 23 |
| 2 | 2 | 40 |
| 1 | 3 | 27 |
| 0 | 4 | 4 |

Does this outcome differ significantly from what would be predicted by chance alone?

14.7 In 120 rolls of 5 dice, the following results were obtained.

| Number of Sixes | Observed Frequency |
|---|---|
| 5 | 6 |
| 4 | 8 |
| 3 | 12 |
| 2 | 20 |
| 1 | 39 |
| 0 | 35 |

Assuming that the dice are "fair," is it likely that this outcome would occur by chance alone? (Or more to the point, would you care to gamble with the owner of these dice?)

14.8 Refer to example 5.4 concerning the number of maple seedlings per quadrat in a random sample of 100 quadrats. Using the sample mean as the best estimate of the parametric mean, determine if the observed results follow a Poisson distribution. Interpret your results biologically.

14.9 Refer to exercise 2.3 concerning the distribution of ant lion pits. Using the sample mean as the best estimate of the parametric mean, determine if the distribution of pits follows a Poisson distribution. Interpret your results biologically.

### Chi-square Test of Association

14.10 It is suspected that female water snakes that forage in Lake Michigan migrate to inland ponds in the fall to deliver their young. If this is correct, one might expect that females would be much more likely to migrate at that time than would males. The following data were collected. (Data from C. Meyers.)

| | Migrators | Nonmigrators |
|---|---|---|
| Females | 25 | 2 |
| Males | 4 | 30 |

Is there an association between sex and migration?

14.11 White-throated sparrows occur in 2 distinct color morphs, referred to as brown and white. It was suspected that females select mates of the opposite morph (i.e., white females select brown males and vice versa). This phenomenon is known as negative assortative mating. In 30 mated pairs, the color combinations were as follows. (Data from D. Tuzzalino.)

| | | Males | |
|---|---|---|---|
| | | White | Brown |
| Females | White | 7 | 23 |
| | Brown | 14 | 5 |

Do the results support the assumption that negative assortative mating occurs in this species?

*14.12 Using the data from digital appendix 3, determine if male university students are more or less likely to smoke than are female university students (i.e., determine if there is an association between smoking and sex).

*14.13 Using the data from digital appendix 3, determine if male university students are more or less likely to do aerobic exercises on a regular basis than are female university students (i.e., determine if there is an association between sex and aerobic exercising).

14.14 The frequency of 3 food items (small snails, cladocerans, and mosquito larvae) in the stomach contents of male and female mosquito fish was determined. Do males and females differ with respect to food items utilized?

| | Food Item | | |
|---|---|---|---|
| | Snail | Cladoceran | Mosquito Larvae |
| Males | 50 | 23 | 15 |
| Females | 10 | 14 | 62 |

14.15 Suppose that the outcome of exercise 14.11 was as follows.

| | | Males | |
|---|---|---|---|
| | | White | Brown |
| Females | White | 1 | 6 |
| | Brown | 5 | 2 |

Determine if the null hypothesis of no association may be rejected. Note that some expected cell sizes are <5.

14.16 An epidemiologist investigating an outbreak of food poisoning (gastroenteritis) at a church picnic used two steps to determine the cause. She first surveyed everyone attending the picnic to determine which food(s) is most likely to have caused the illness. (She later confirmed the cause by sampling food and ill people and culturing the bacteria.) To survey, she asked each person whether he or she became ill (vomiting and/or diarrhea within 24 hours of the picnic) and specifically which foods he or she ate and did not eat. Compiling information from 66 respondents provided the following data (from Lilienfeld and Lilienfeld, 1980). Use a chi-square test of independence to determine if each of the foods is linked to illness. Note that you will not be able to test the first food (chicken) because only one person did not eat the chicken.

| | Ill | Not Ill | Total | Attack rate (%) |
|---|---|---|---|---|
| Chicken: | | | | |
| Eaten | 49 | 16 | 65 | 75 |
| Not Eaten | 0 | 1 | 1 | 0 |
| | | | difference | 75 |
| Chicken Dressing: | | | | |
| Eaten | 37 | 5 | 42 | 88 |
| Not Eaten | 11 | 11 | 22 | 50 |
| | | | difference | 38 |
| Potato Salad: | | | | |
| Eaten | 37 | 5 | 42 | 88 |
| Not Eaten | 11 | 9 | 20 | 55 |
| | | | difference | 33 |
| Cole Slaw: | | | | |
| Eaten | 27 | 7 | 34 | 79 |
| Not Eaten | 17 | 9 | 26 | 65 |
| | | | difference | 14 |

14.17 An article in the *Springfield News Leader* (11/7/03) summarized results from a British study that concluded that marijuana (cannabis) relieves symptoms of multiple sclerosis (MS). The study stated that 630 MS patients were included in an experiment in which they were randomly assigned to one of three treatments, with equal numbers in each treatment. The treatments were: cannabis oil, synthetic cannabis (THC), and a placebo control. This was a double blind study (i.e., neither patients nor their doctors know which treatment they received). The following numbers are *percentages* of each group that experienced symptom relief. For each disease symptom, test the null hypothesis that treatment and symptom relief are independent. Do the results support the conclusions in the article?

| Treatment | Pain relieved (%) | Stiffness relieved(%) |
|---|---|---|
| Cannabis | 57 | 61 |
| THC | 50 | 60 |
| Placebo | 37 | 46 |

14.18 Locate a set of frequency data either from the popular press or from government web sites, such as the Centers for Disease Control (http://www.cdc.gov/mmwr/). Generate a contingency table. Next, state the hypothesis you are testing. Then, conduct a chi-square test, reporting both the chi-square statistic and *p* value, make a decision about your hypothesis, and report a verbal conclusion.

### Choose the Correct Test

14.19 In order to study the dependence of hypertension on smoking habits, the following data were collected from 180 individuals. Test the hypothesis that the presence or absence of hypertension is independent of smoking habits.

| | Smoking habit | | |
|---|---|---|---|
| | non- | moderate | heavy |
| Hypertension | 21 | 36 | 30 |
| No hypertension | 28 | 26 | 19 |

14.20 In snapdragons, red flower color is incompletely dominant. Homozygous dominant individuals are red, heterozygous individuals are pink, and homozygous recessive individuals are white. In a cross of 2 heterozygous individuals, a ratio of 1 red to 2 pink to 1 white is expected in the offspring. The results of such a cross were 10 red, 21 pink, and 9 white. Do the observed results differ significantly from a 1:2:1 ratio?

14.21 Forty-seven groups of common suckers, each consisting of 3 individuals, were surveyed during their spawning season. The sex ratio in the population may be assumed to be 1:1. The number of males and females in each group of 3 were as follows.

| Males | Females | Frequency |
|---|---|---|
| 3 | 0 | 7 |
| 2 | 1 | 35 |
| 1 | 2 | 3 |
| 0 | 3 | 2 |

Is it likely that the number of males and females in groups of 3 individuals are due to chance alone?

14.22 Based on our understanding of X and Y chromosomes, it is expected that female births equal male births in most mammalian species. In a sample of 60 individuals from a population, 28 were males, and 32 were females. May we conclude that the sex ratio in this population is something other than 1:1?

14.23 As part of a conservation-biology study, a researcher used radio transmitters to locate winter roosts of the red bat (Lasiurus borealis). The following data show the number of trees containing roosts compared with randomly chosen trees that lacked roosts. May we conclude that there is an association between bat roosts and tree species? (Data from B. Mormann.)

| Tree species | With roosts | Without roost |
|---|---|---|
| Cedar (*Juniperus*) | 11 | 6 |
| Oak (*Quercus*) | 10 | 15 |
| Other | 0 | 21 |

# 15

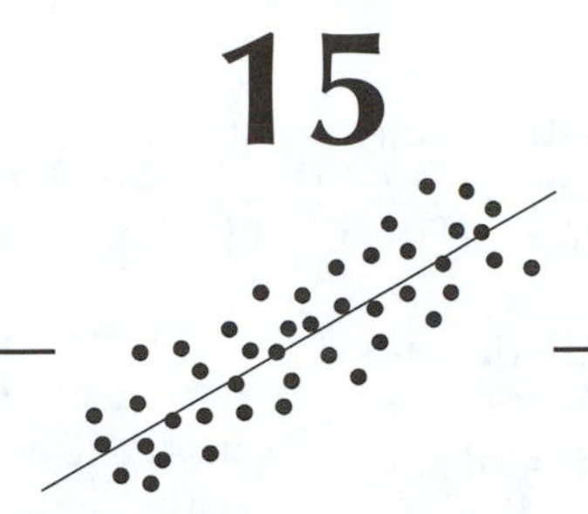

# Choice of Tests and a View of Some Other Procedures

If you are like most students of biostatistics, you probably feel like you have "gone through the wringer." The amount of numerical data statistics deals with sometimes makes it seem less like biology and more like math. But we must not lose sight of the primary purpose of this ordeal—to use data to make inferences about interesting questions. (And biologists have lots of questions!) Statistics allows us to make inferences more efficiently and in a way that other scientists generally accept. In this book, you have been introduced to a variety of statistical procedures that are appropriate for different sorts of data. In this final chapter, we first review the general approach for selecting the correct statistical test and use a simple schematic drawing to act like a key. We then review some general principles of experimental design. Finally, we briefly explore some newer and specialized statistical procedures and where one can go for further information.

## 15.1 Choice of the Appropriate Statistical Test

Recall that the selection of a statistical test depends on a number of things. What is the hypothesis of interest? What type of variable(s) were measured or counted (section 2.1)? Be able to list and describe the variables and, if categorical, how many different levels of each were selected. What is the structure of the data? Specifically, how many populations were sampled? Are samples related (sections 9.5 and 11.1)? Are the assumptions of parametric statistics met (section 10.4)? What are your sample sizes? Even if you were not doing the analysis yourself, but were instead going to a statistical consultant, (s)he would ask for these kinds of information.

We can use a simple schematic device to aid the selection of statistical tests. Figure 15.1 illustrates a key for selecting tests with which you are now familiar. Recall that we used a shorter version of this key after we covered $t$-tests in chapter 9. This schematic asks many of the questions in the previous paragraph in order to choose the right test. Most often, you will need to read further in a statistics book about the particular test and check examples to be sure that test applies to your data set. Please keep in mind that this listing is not exhaustive, as there are other types of statistics that we have not covered in this book. Brief descriptions of some of these procedures are given below in section 15.2 and further information appears in the references.

Recall the general purposes for which we have used statistics. Descriptive statistics summarize large masses of data into simpler morsels more easily digested by the researcher or the reader. Inferential statistics are used for hypothesis testing (the bulk of this book). Statistics can also be used to guide experimental design, improving our ability to design meaningful and efficient studies.

## 15.2 Experimental Design

In section 9.3, we explore one aspect of experimental design, determining the sample size needed to detect a minimum effect size. Other design issues come up as well, such as the importance of independence among data (sections 1.4 and 14.3). Experimental design is presented thoroughly in several outstanding references (Box et al. 1978, Mead 1992, Quinn and Keough 2002). Green (1979), in his book on statistics for environmental biologists, provides an excellent summary of principles for good experimental design, which all users of statistics should heed. Many of these principles should be quite clear to you by now. Those principles are summarized below.

## 15.3 A View of Some Other Statistical Procedures

With a limitation of space (and your patience!), we have had to focus our attention in this text on only the most basic and widely used statistical procedures. A brief description of several other procedures should provide a glimpse of some new possibilities.

**Multivariate statistics** are widely used in the fields of ecology and systematics, as well as in the social sciences. These procedures reduce a larger number of response variables into a smaller set of variables that capture the most important features of variation in the data set and can be shown in plots. Examples of multivariate procedures widely used in biology include principal components analysis (PCA), canonical correspondence analysis (CCA), multivariate ANOVA, discriminant function analysis, and cluster analysis. For descriptions and examples of these tests, see Stevens (1996), McGarigal et al. (2000), or one of the many other books published on multivariate statistics. Most widely used computer statistics packages (such as MINITAB, SPSS, SAS) include programs for multivariate analysis, although not all packages contain all the procedures. Several packages are dedicated solely to multivariate analysis (e.g., CANOCO, PC-ORD).

**Ten Principles for Sound Experimental Design (Modified after Green, 1979)**

1. Clearly establish the *hypothesis* being tested and keep the experiment focused on that hypothesis. Be able to explain to someone else the question you are asking.
2. Hypotheses may be tested either with manipulative *experiments* or carefully focused *surveys*. However, cause and effect is easier to establish with experiments.
3. Keep the design as simple as possible. Generally, only a single hypothesis should be addressed with each study. (Exceptions are specific designs, such as two-way ANOVA.)
4. Plan to collect *numerical data*. These data may consist of measurements, ranks, or counts of attributes (frequencies).
5. Include a *control*. To test whether a condition has an effect, collect samples both where the condition is present and where the condition is absent. These different *treatment* groups should be alike except for the variable you are trying to test.
6. Use *replication*. Collect multiple measurements from each treatment group. The required sample size depends on underlying variation.
7. Use *randomization*. Assignment of subjects to different experimental treatments should be done at random, by use of either a random-number table or a computer. (Proper randomization also helps assure *independence*.)
8. Beware of *bias*. There is a natural tendency to record what we expect to see, rather than what is actually there. (The double-blind study is an example of an experimental procedure to reduce bias.)
9. Illustrate your data with simple tables and graphs that show both averages and variation. Differences between groups can be interpreted only in light of variation within groups.
10. Statistical tests can be used to determine whether the differences between groups are large enough to consider them truly different.

Figure 15.1 A key to commonly-used statistical tests. Modified after Ambrose and Ambrose (1981)

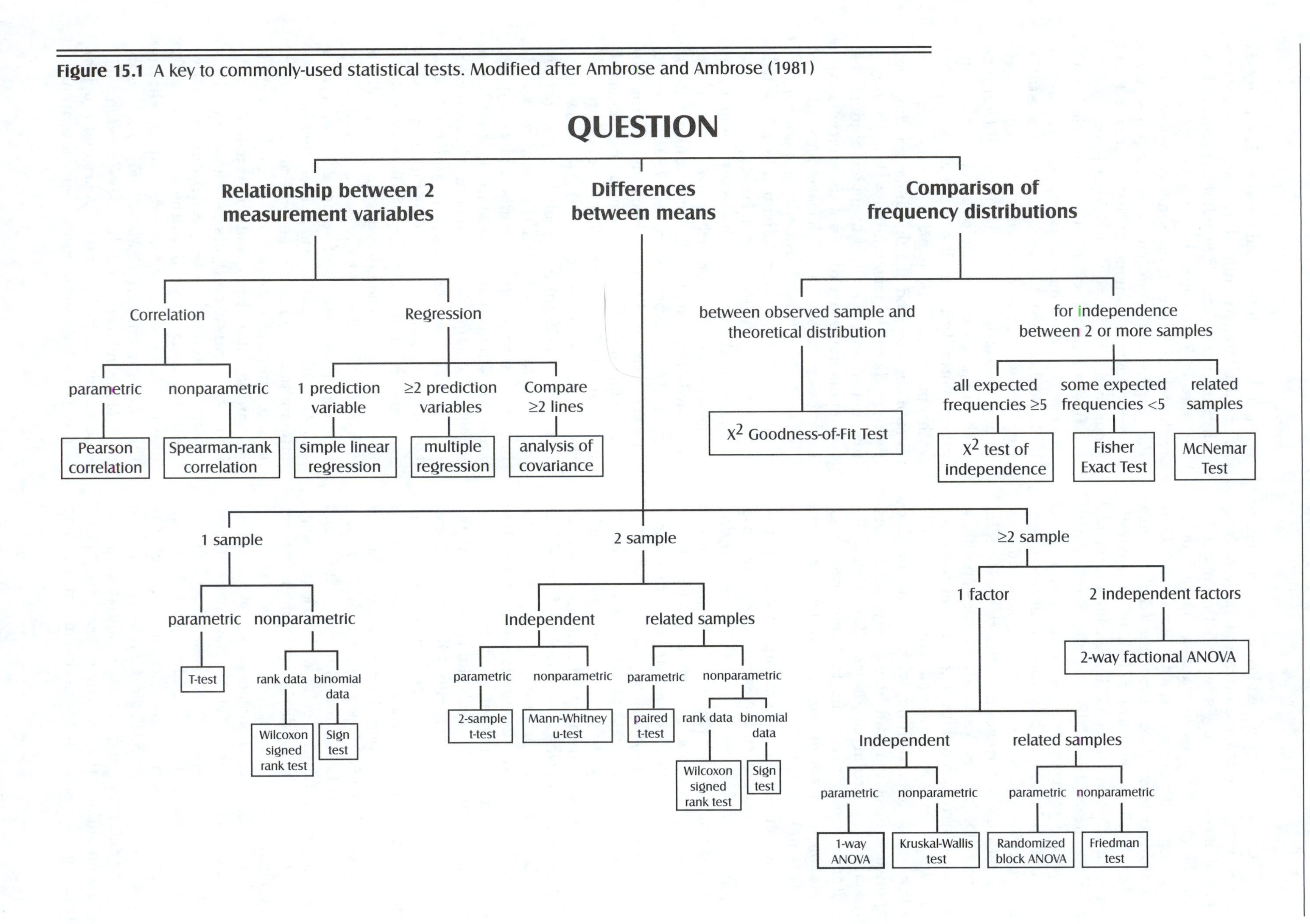

With the current explosion in computing power, graphic-intensive programs have recently become possible (as any science-fiction movie fan would know). This use of graphics has had important implications for spatial analysis. **Geographic information systems** (GIS) is now widely used to organize and display spatially structured data. Hence GIS is an important tool for data description. GIS also allows calculation of a variety of geographic measures that are used for many purposes in business, government, and science. For instance, stream ecologists and planners both use information on the risk of flooding in river floodplains. Various statistics specific to spatially structured data (e.g., Mantel tests) are presented in Scheiner and Gurevitch (2001). Other questions dealing with frequency data applied to direction (and time of day) can be analyzed with circular statistics (Zar 1999). For instance, we might want to know from what direction most storms come or which directions birds tend to migrate.

Another consequence of improved computing power is the evolution of computer-intensive resampling tests. Examples include **bootstrapping** and jackknife tests. These are especially useful where large amounts of effort go into obtaining a single estimate and we wish to determine confidence intervals around that estimate. For example, a life table consists of age-structured survivorship and mortality data, which can be used by ecologists to generate a single estimate of population growth rate, which we'll call *r* ("fitness"). To do a bootstrap, we program a computer to randomly discard a single individual from the data set, use the remaining information to compute *r*, and then replace the individual and repeat the process many times (typically 1000 or more iterations). All these computer-generated *r* values are then used to compute descriptive statistics (mean and standard deviation) for *r* and, from these, a confidence interval. Resampling techniques are introduced in Scheiner and Gurevitch (2001) and the larger statistical software packages (e.g., MINITAB) can be readily programmed to carry them out.

Many biological phenomena show patterning in time. For instance, female hormones change with a monthly cycle and the behaviors of numerous creatures vary over a 24-hour period. Detecting the pattern is sometimes easy (graph the data!). Certain physiological cycles are not so easy to detect and require use of statistical procedures to pull out the chief signal from the "noise" (random variation). If the period of study is long enough, a procedure called **time series analysis** can be used to determine if the cycling is statistically discernable and, if so, what the period of the response is. This method is also applied to long-term studies of the climate and economic cycles. For further information, see statistics texts directed toward economists (e.g., Wonnacott and Wonnacott 1977) for detailed descriptions.

Biostatistics is widely used in public health. Besides the traditional statistics covered in this book, public-health workers explore **vital statistics** (births, deaths, etc.), which are kept as government records and can be used to determine such things as age- specific mortality rates and infant mortality rates. Biostatisticians, together with epidemiologists, explore factors related to human disease and death. Understanding patterns of disease in populations through survey techniques provides clues to understanding risk factors in circulatory diseases and cancer, as well as tracking the causes of disease outbreaks and epidemics. The role of statistics in public health is described in detail in Lilienfeld and Lilienfeld (1980) and Forthofer and Lee (1995).

Finally, statistics can be used to guide data reviews in areas where extensive research has previously been conducted. For instance, is there a general consensus that compound X is linked to cancer? Occasionally the public is misinformed by the media on conflicts between the results of different scientific studies. (The media, after all, thrives on perceived conflict!) We know from basic design principles (section 15.1) that certain key mistakes can invalidate the results of a study (e.g., failure to collect random samples). We also know from the principles of statistics that larger sample size improves estimation of parameters (section 7.3) and the power of statistical tests (section 9.3). Therefore, studies differing in sample sizes can differ in their information content. Also, we may occasionally make a mistake about a hypothesis, even with a well-designed study, because of the nature of samples (recall type I and type II errors). Formal literature reviews attempt to determine the general consensus from a large number of studies. Poorly designed studies can be filtered out and the remainder subjected to a **meta-analysis**. A meta-analysis is essentially the process of generating statistics about statistics. Overall, is there a statistically discernable effect of compound X? If so, what is the minimum effect size? Meta-analysis allows us to explore subjects that have been extensively studied and argued for many years and provides a way to contribute toward a general consensus.

An entry into literature on meta-analysis can be found in Scheiner and Gurevitch (2001).

These other statistical tools provide exciting new opportunities for exploring biological data and they are becoming part of our standard vocabulary in statistics. Overall, they have the same usage as the other procedures studied in this book—data description, hypothesis testing, and guiding experimental design.

## Key Terms

bootstrapping
experimental design
geographic information systems (GIS)
meta-analysis
multivariate statistics
time series analysis
vital statistics

# Appendix A

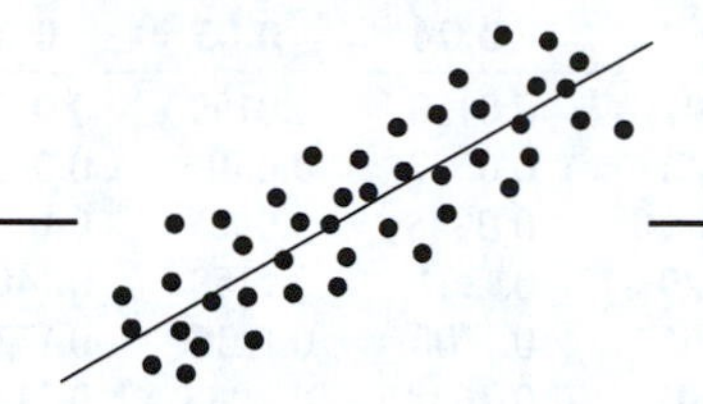

# Statistical Tables

**Table A.1** Areas of the normal distribution

| z | 0.00 | 0.01 | 0.02 | 0.03 | 0.04 | 0.05 | 0.06 | 0.07 | 0.08 | 0.09 |
|---|---|---|---|---|---|---|---|---|---|---|
| 0.0 | 0.0000 | 0.0040 | 0.0080 | 0.0120 | 0.0160 | 0.0199 | 0.0239 | 0.0279 | 0.0319 | 0.0359 |
| 0.1 | 0.0398 | 0.0438 | 0.0478 | 0.0517 | 0.0557 | 0.0596 | 0.0636 | 0.0675 | 0.0714 | 0.0754 |
| 0.2 | 0.0793 | 0.0832 | 0.0871 | 0.0910 | 0.0948 | 0.0987 | 0.1026 | 0.1064 | 0.1103 | 0.1141 |
| 0.3 | 0.1179 | 0.1217 | 0.1255 | 0.1293 | 0.1331 | 0.1368 | 0.1406 | 0.1443 | 0.1480 | 0.1517 |
| 0.4 | 0.1554 | 0.1591 | 0.1628 | 0.1664 | 0.1700 | 0.1736 | 0.1772 | 0.1808 | 0.1844 | 0.1879 |
| 0.5 | 0.1915 | 0.1950 | 0.1985 | 0.2019 | 0.2054 | 0.2088 | 0.2123 | 0.2157 | 0.2190 | 0.2224 |
| 0.6 | 0.2258 | 0.2291 | 0.2324 | 0.2357 | 0.2389 | 0.2422 | 0.2454 | 0.2486 | 0.2518 | 0.2549 |
| 0.7 | 0.2580 | 0.2612 | 0.2642 | 0.2673 | 0.2704 | 0.2734 | 0.2764 | 0.2794 | 0.2823 | 0.2852 |
| 0.8 | 0.2881 | 0.2910 | 0.2939 | 0.2967 | 0.2996 | 0.3023 | 0.3051 | 0.3078 | 0.3106 | 0.3133 |
| 0.9 | 0.3159 | 0.3186 | 0.3212 | 0.3238 | 0.3264 | 0.3289 | 0.3315 | 0.3340 | 0.3365 | 0.3389 |
| 1.0 | 0.3413 | 0.3438 | 0.3461 | 0.3485 | 0.3508 | 0.3531 | 0.3554 | 0.3577 | 0.3599 | 0.3621 |
| 1.1 | 0.3643 | 0.3665 | 0.3686 | 0.3708 | 0.3729 | 0.3749 | 0.3770 | 0.3790 | 0.3810 | 0.3830 |
| 1.2 | 0.3849 | 0.3869 | 0.3888 | 0.3907 | 0.3925 | 0.3944 | 0.3962 | 0.3980 | 0.3997 | 0.4015 |
| 1.3 | 0.4032 | 0.4049 | 0.4066 | 0.4082 | 0.4099 | 0.4115 | 0.4131 | 0.4147 | 0.4162 | 0.4177 |
| 1.4 | 0.4192 | 0.4207 | 0.4222 | 0.4236 | 0.4251 | 0.4265 | 0.4279 | 0.4292 | 0.4306 | 0.4319 |
| 1.5 | 0.4332 | 0.4345 | 0.4357 | 0.4370 | 0.4382 | 0.4394 | 0.4406 | 0.4418 | 0.4429 | 0.4441 |
| 1.6 | 0.4452 | 0.4463 | 0.4474 | 0.4484 | 0.4495 | 0.4505 | 0.4515 | 0.4525 | 0.4535 | 0.4545 |
| 1.7 | 0.4554 | 0.4564 | 0.4573 | 0.4582 | 0.4591 | 0.4599 | 0.4608 | 0.4616 | 0.4625 | 0.4633 |
| 1.8 | 0.4641 | 0.4649 | 0.4656 | 0.4664 | 0.4671 | 0.4678 | 0.4686 | 0.4693 | 0.4699 | 0.4706 |
| 1.9 | 0.4713 | 0.4719 | 0.4726 | 0.4732 | 0.4738 | 0.4744 | 0.4750 | 0.4756 | 0.4761 | 0.4767 |
| 2.0 | 0.4772 | 0.4778 | 0.4783 | 0.4788 | 0.4793 | 0.4798 | 0.4803 | 0.4808 | 0.4812 | 0.4817 |
| 2.1 | 0.4821 | 0.4826 | 0.4830 | 0.4834 | 0.4838 | 0.4842 | 0.4846 | 0.4850 | 0.4854 | 0.4857 |
| 2.2 | 0.4861 | 0.4864 | 0.4868 | 0.4871 | 0.4875 | 0.4878 | 0.4881 | 0.4884 | 0.4887 | 0.4890 |
| 2.3 | 0.4893 | 0.4896 | 0.4898 | 0.4901 | 0.4904 | 0.4906 | 0.4909 | 0.4911 | 0.4913 | 0.4916 |
| 2.4 | 0.4918 | 0.4920 | 0.4922 | 0.4925 | 0.4927 | 0.4929 | 0.4931 | 0.4932 | 0.4934 | 0.4936 |
| 2.5 | 0.4938 | 0.4940 | 0.4941 | 0.4943 | 0.4945 | 0.4946 | 0.4948 | 0.4949 | 0.4951 | 0.4952 |
| 2.6 | 0.4953 | 0.4955 | 0.4956 | 0.4957 | 0.4959 | 0.4960 | 0.4961 | 0.4962 | 0.4963 | 0.4964 |
| 2.7 | 0.4965 | 0.4966 | 0.4967 | 0.4968 | 0.4969 | 0.4970 | 0.4971 | 0.4972 | 0.4973 | 0.4974 |
| 2.8 | 0.4974 | 0.4975 | 0.4976 | 0.4977 | 0.4977 | 0.4978 | 0.4979 | 0.4979 | 0.4980 | 0.4981 |
| 2.9 | 0.4981 | 0.4982 | 0.4982 | 0.4983 | 0.4984 | 0.4984 | 0.4985 | 0.4985 | 0.4986 | 0.4986 |
| 3.0 | 0.4987 | 0.4987 | 0.4987 | 0.4988 | 0.4988 | 0.4989 | 0.4989 | 0.4989 | 0.4990 | 0.4990 |
| 3.1 | 0.4990 | 0.4991 | 0.4991 | 0.4991 | 0.4992 | 0.4992 | 0.4992 | 0.4992 | 0.4993 | 0.4993 |
| 3.2 | 0.4993 | 0.4993 | 0.4994 | 0.4994 | 0.4994 | 0.4994 | 0.4994 | 0.4995 | 0.4995 | 0.4995 |
| 3.3 | 0.4995 | 0.4995 | 0.4995 | 0.4996 | 0.4996 | 0.4996 | 0.4996 | 0.4996 | 0.4996 | 0.4997 |
| 3.4 | 0.4997 | 0.4997 | 0.4997 | 0.4997 | 0.4997 | 0.4997 | 0.4997 | 0.4997 | 0.4997 | 0.4998 |
| 3.5 | 0.4998 | 0.4998 | 0.4998 | 0.4998 | 0.4998 | 0.4998 | 0.4998 | 0.4998 | 0.4998 | 0.4998 |
| 3.6 | 0.4998 | 0.4998 | 0.4999 | 0.4999 | 0.4999 | 0.4999 | 0.4999 | 0.4999 | 0.4999 | 0.4999 |
| 3.7 | 0.4999 | 0.4999 | 0.4999 | 0.4999 | 0.4999 | 0.4999 | 0.4999 | 0.4999 | 0.4999 | 0.4999 |
| 3.8 | 0.4999 | 0.4999 | 0.4999 | 0.4999 | 0.4999 | 0.4999 | 0.4999 | 0.4999 | 0.4999 | 0.4999 |
| 3.9 | 0.49995 | 0.49995 | 0.49996 | 0.49996 | 0.49996 | 0.49996 | 0.49996 | 0.49996 | 0.49997 | 0.49997 |

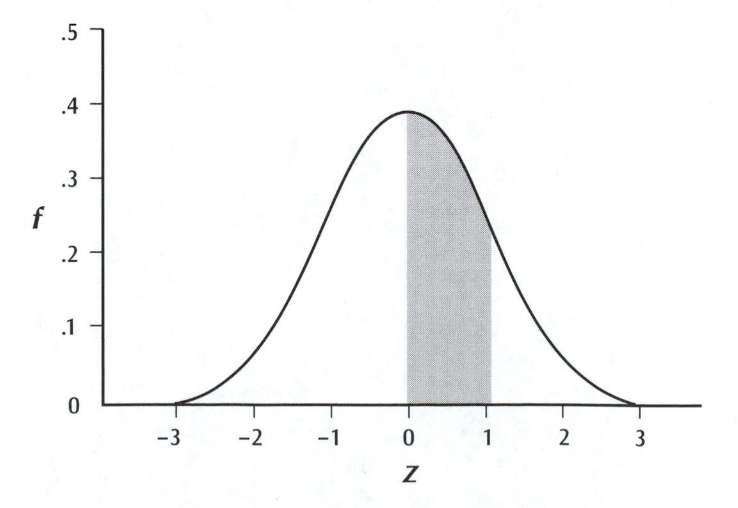

**Table A.2** Critical values of the *t* distribution

| df | $\alpha$ (Two-Tailed) 0.2 | 0.1 | 0.05 | 0.02 | 0.01 | 0.001 | 0.0001 |
|---|---|---|---|---|---|---|---|
| 1 | 3.078 | 6.314 | 12.706 | 31.821 | 63.657 | 636.619 | 6366.198 |
| 2 | 1.886 | 2.920 | 4.303 | 6.695 | 9.925 | 31.598 | 99.992 |
| 3 | 1.638 | 2.353 | 3.182 | 4.541 | 5.841 | 12.924 | 28.000 |
| 4 | 1.533 | 2.132 | 2.776 | 3.747 | 4.604 | 8.610 | 15.544 |
| 5 | 1.476 | 2.015 | 2.571 | 3.365 | 4.032 | 6.869 | 11.178 |
| 6 | 1.44 | 1.943 | 2.447 | 3.143 | 3.707 | 5.959 | 9.082 |
| 7 | 1.415 | 1.895 | 2.365 | 2.998 | 3.499 | 5.408 | 7.885 |
| 8 | 1.397 | 1.860 | 2.306 | 2.896 | 3.355 | 5.041 | 7.120 |
| 9 | 1.383 | 1.833 | 2.262 | 2.821 | 3.250 | 4.781 | 6.594 |
| 10 | 1.372 | 1.812 | 2.228 | 2.764 | 3.169 | 4.587 | 6.211 |
| 11 | 1.363 | 1.796 | 2.201 | 2.718 | 3.106 | 4.437 | 5.921 |
| 12 | 1.356 | 1.782 | 2.179 | 2.681 | 3.055 | 4.318 | 5.694 |
| 13 | 1.35 | 1.771 | 2.160 | 2.650 | 3.012 | 4.221 | 5.513 |
| 14 | 1.345 | 1.761 | 2.145 | 2.624 | 2.977 | 4.140 | 5.363 |
| 15 | 1.341 | 1.753 | 2.131 | 2.602 | 2.947 | 4.073 | 5.239 |
| 16 | 1.337 | 1.746 | 2.120 | 2.583 | 2.921 | 4.015 | 5.134 |
| 17 | 1.333 | 1.740 | 2.110 | 2.567 | 2.898 | 3.965 | 5.044 |
| 18 | 1.33 | 1.734 | 2.101 | 2.552 | 2.878 | 3.922 | 4.966 |
| 19 | 1.328 | 1.729 | 2.093 | 2.539 | 2.861 | 3.883 | 4.897 |
| 20 | 1.325 | 1.725 | 2.086 | 2.528 | 2.845 | 3.850 | 4.837 |
| 21 | 1.323 | 1.721 | 2.080 | 2.518 | 2.831 | 3.819 | 4.784 |
| 22 | 1.321 | 1.717 | 2.074 | 2.508 | 2.819 | 3.792 | 4.736 |
| 23 | 1.319 | 1.714 | 2.069 | 2.500 | 2.807 | 3.767 | 4.693 |
| 24 | 1.318 | 1.711 | 2.064 | 2.492 | 2.797 | 3.745 | 4.654 |
| 25 | 1.316 | 1.708 | 2.060 | 2.485 | 2.787 | 3.725 | 4.619 |
| 26 | 1.315 | 1.706 | 2.056 | 2.479 | 2.779 | 3.707 | 4.587 |
| 27 | 1.314 | 1.703 | 2.052 | 2.473 | 2.771 | 3.690 | 4.558 |
| 28 | 1.313 | 1.701 | 2.048 | 2.467 | 2.763 | 3.674 | 4.530 |
| 29 | 1.311 | 1.699 | 2.045 | 2.462 | 2.756 | 3.659 | 4.506 |
| 30 | 1.31 | 1.697 | 2.042 | 2.457 | 2.750 | 3.646 | 4.482 |
| 40 | 1.303 | 1.684 | 2.021 | 2.423 | 2.704 | 3.551 | 4.321 |
| 60 | 1.296 | 1.671 | 2.000 | 2.390 | 2.660 | 3.460 | 4.169 |
| 100 | 1.292 | 1.660 | 1.984 | 2.364 | 2.626 | 3.390 | 4.053 |
| ∞ | 1.282 | 1.645 | 1.960 | 2.326 | 2.576 | 3.291 | 3.750 |

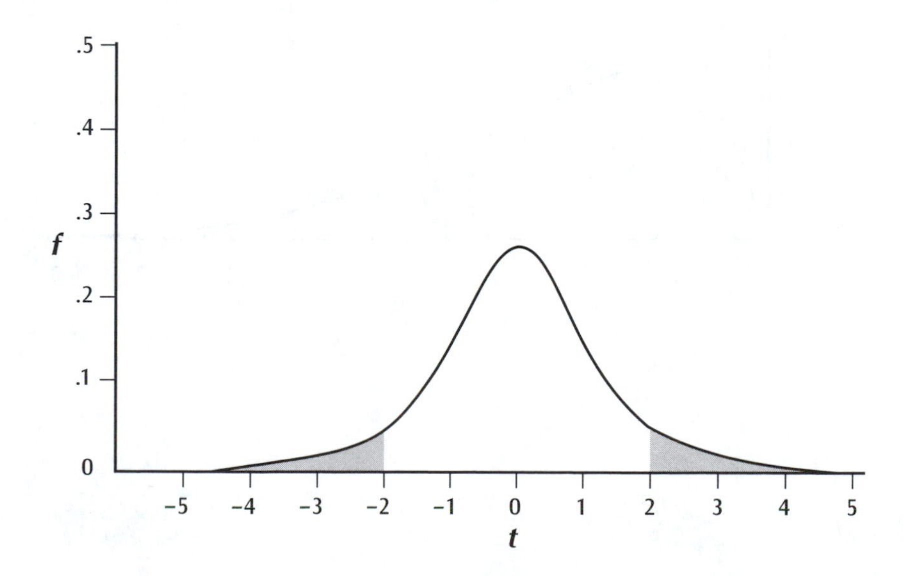

**Table A.3** Critical values of the chi-square distribution

| df | $\alpha$ 0.05 | 0.02 | 0.01 | 0.001 | 0.0001 |
|---|---|---|---|---|---|
| 1 | 3.84 | 5.41 | 6.63 | 10.83 | 15.14 |
| 2 | 5.99 | 7.82 | 9.21 | 13.82 | 18.42 |
| 3 | 7.81 | 9.84 | 11.34 | 16.27 | 21.11 |
| 4 | 9.49 | 11.67 | 13.28 | 18.47 | 23.51 |
| 5 | 11.07 | 13.39 | 15.09 | 20.51 | 25.74 |
| 6 | 12.59 | 15.03 | 16.81 | 22.46 | 27.86 |
| 7 | 14.07 | 16.62 | 18.48 | 24.32 | 29.88 |
| 8 | 15.51 | 18.17 | 20.09 | 26.12 | 31.83 |
| 9 | 16.92 | 19.68 | 21.67 | 27.88 | 33.72 |
| 10 | 18.31 | 21.16 | 23.21 | 29.59 | 35.56 |
| 11 | 19.68 | 22.62 | 24.72 | 31.26 | 37.37 |
| 12 | 21.03 | 24.05 | 26.22 | 32.91 | 39.13 |
| 13 | 22.36 | 25.47 | 27.69 | 34.53 | 40.87 |
| 14 | 23.68 | 26.87 | 29.14 | 36.12 | 42.58 |
| 15 | 25.00 | 28.26 | 30.58 | 37.70 | 44.26 |
| 16 | 26.30 | 29.63 | 32.00 | 39.25 | 45.92 |
| 17 | 27.59 | 31.00 | 33.41 | 40.79 | 47.57 |
| 18 | 28.87 | 32.35 | 34.81 | 42.31 | 49.19 |
| 19 | 30.14 | 33.69 | 36.19 | 43.82 | 50.80 |
| 20 | 31.41 | 35.02 | 37.57 | 45.31 | 52.39 |
| 21 | 32.67 | 36.34 | 38.93 | 46.80 | 53.96 |
| 22 | 33.92 | 37.66 | 40.29 | 48.27 | 55.52 |
| 23 | 35.17 | 38.97 | 41.64 | 49.73 | 57.08 |
| 24 | 36.42 | 40.27 | 42.98 | 51.18 | 58.61 |
| 25 | 37.65 | 41.57 | 44.31 | 52.62 | 60.14 |
| 26 | 38.89 | 42.86 | 45.64 | 54.05 | 61.66 |
| 27 | 40.11 | 44.14 | 46.96 | 55.48 | 63.16 |
| 28 | 41.34 | 45.42 | 48.28 | 56.89 | 64.66 |
| 29 | 42.56 | 46.69 | 49.59 | 58.30 | 66.15 |
| 30 | 43.77 | 47.96 | 50.89 | 59.70 | 67.63 |

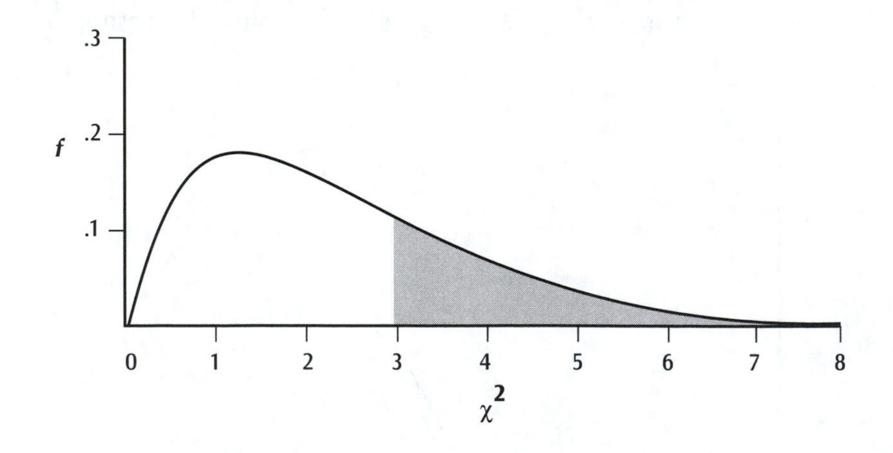

**Table A.4** Critical values of $U$ for the Mann-Whitney test. Sample sizes for each group are denoted by $n_1$ and $n_2$ (it does not matter which group is which). At $\alpha = 0.05$, use the top table for a two-tailed test and bottom table for a one-tailed test. (The top table may also be used for a one-tailed test at $\alpha = 0.025$, and the bottom table for a two-tailed test at $\alpha = 0.10$)

**Two-tailed test**

| | | | | | | | | $n_1$ | | | | | | | | |
|---|---|---|---|---|---|---|---|---|---|---|---|---|---|---|---|---|
| **5** | **6** | **7** | **8** | **9** | **10** | **11** | **12** | **13** | **14** | **15** | **16** | **17** | **18** | **19** | **20** | $n_2$ |
| 23 | 27 | 30 | 34 | 38 | 42 | 46 | 49 | 53 | 57 | 61 | 65 | 68 | 72 | 76 | 80 | 5 |
| | 31 | 36 | 40 | 44 | 49 | 53 | 58 | 62 | 67 | 71 | 75 | 80 | 84 | 89 | 93 | 6 |
| | | 41 | 46 | 51 | 56 | 61 | 66 | 71 | 76 | 81 | 86 | 91 | 96 | 101 | 106 | 7 |
| | | | 51 | 57 | 63 | 69 | 74 | 80 | 86 | 91 | 97 | 102 | 108 | 114 | 119 | 8 |
| | | | | 64 | 70 | 76 | 82 | 89 | 95 | 101 | 107 | 114 | 120 | 126 | 132 | 9 |
| | | | | | 77 | 84 | 91 | 97 | 104 | 111 | 118 | 125 | 132 | 138 | 145 | 10 |
| | | | | | | 91 | 99 | 106 | 114 | 121 | 129 | 136 | 143 | 151 | 158 | 11 |
| | | | | | | | 107 | 115 | 123 | 131 | 139 | 147 | 155 | 163 | 171 | 12 |
| | | | | | | | | 124 | 132 | 141 | 149 | 158 | 167 | 175 | 184 | 13 |
| | | | | | | | | | 141 | 151 | 160 | 169 | 178 | 188 | 197 | 14 |
| | | | | | | | | | | 161 | 170 | 180 | 190 | 200 | 210 | 15 |
| | | | | | | | | | | | 181 | 191 | 202 | 212 | 222 | 16 |
| | | | | | | | | | | | | 202 | 213 | 224 | 235 | 17 |
| | | | | | | | | | | | | | 225 | 236 | 248 | 18 |
| | | | | | | | | | | | | | | 248 | 261 | 19 |
| | | | | | | | | | | | | | | | 273 | 20 |

**One-tailed test**

| | | | | | | | | $n_1$ | | | | | | | | |
|---|---|---|---|---|---|---|---|---|---|---|---|---|---|---|---|---|
| **5** | **6** | **7** | **8** | **9** | **10** | **11** | **12** | **13** | **14** | **15** | **16** | **17** | **18** | **19** | **20** | $n_2$ |
| 21 | | | | | | | | | | | | | | | | 5 |
| 25 | 29 | | | | | | | | | | | | | | | 6 |
| 29 | 34 | 38 | | | | | | | | | | | | | | 7 |
| 32 | 38 | 43 | 49 | | | | | | | | | | | | | 8 |
| 36 | 42 | 48 | 54 | 60 | | | | | | | | | | | | 9 |
| 39 | 46 | 53 | 60 | 66 | 73 | | | | | | | | | | | 10 |
| 43 | 50 | 58 | 65 | 72 | 79 | 87 | | | | | | | | | | 11 |
| 47 | 55 | 63 | 70 | 78 | 86 | 94 | 102 | | | | | | | | | 12 |
| 50 | 59 | 67 | 76 | 84 | 93 | 101 | 109 | 118 | | | | | | | | 13 |
| 54 | 63 | 72 | 81 | 90 | 99 | 108 | 117 | 126 | 135 | | | | | | | 14 |
| 57 | 67 | 77 | 87 | 96 | 106 | 115 | 125 | 134 | 144 | 153 | | | | | | 15 |
| 61 | 71 | 82 | 92 | 102 | 112 | 122 | 132 | 143 | 153 | 163 | 173 | | | | | 16 |
| 65 | 76 | 86 | 97 | 108 | 119 | 130 | 140 | 151 | 161 | 172 | 183 | 193 | | | | 17 |
| 68 | 80 | 91 | 103 | 114 | 125 | 137 | 148 | 159 | 170 | 182 | 193 | 204 | 215 | | | 18 |
| 72 | 84 | 96 | 108 | 120 | 132 | 144 | 156 | 167 | 179 | 191 | 203 | 214 | 226 | 238 | | 19 |
| 75 | 88 | 101 | 113 | 126 | 138 | 151 | 163 | 176 | 188 | 200 | 213 | 225 | 237 | 250 | 262 | 20 |

**Table A.5** Critical values of T for the Wilcoxon test. The row *n* represents the number of matched pairs without ties. Reject $H_0$ if the statistic is *less than* the critical value

| | $\alpha$ (2 tailed) | |
|---|---|---|
| *n* | **0.05** | **0.01** |
| 6 | 0 | — |
| 7 | 2 | — |
| 8 | 4 | 0 |
| 9 | 6 | 2 |
| 10 | 8 | 3 |
| 11 | 11 | 5 |
| 12 | 14 | 7 |
| 13 | 17 | 10 |
| 14 | 21 | 13 |
| 15 | 25 | 16 |
| 16 | 30 | 20 |
| 17 | 35 | 23 |
| 18 | 40 | 28 |
| 19 | 46 | 32 |
| 20 | 52 | 38 |
| 21 | 59 | 43 |
| 22 | 66 | 49 |
| 23 | 73 | 55 |
| 24 | 81 | 61 |
| 25 | 89 | 68 |

**Table A.6** Critical values of the *F* distribution ($\alpha = 0.05$; $df_1$ = treatment degrees of freedom, $df_2$ = error degrees of freedom). Part A: $\alpha = 0.05$

| | $df_1$ | | | | | | | | | | | |
|---|---|---|---|---|---|---|---|---|---|---|---|---|
| $df_2$ | 1 | 2 | 3 | 4 | 5 | 6 | 7 | 8 | 9 | 10 | 11 | 12 |
| 2 | 18.5 | 19.0 | 19.2 | 19.3 | 19.4 | 19.4 | 19.4 | 19.4 | 19.4 | 19.4 | 19.4 | 19.4 |
| 3 | 10.1 | 9.55 | 9.28 | 9.12 | 9.01 | 8.94 | 8.89 | 8.85 | 8.81 | 8.79 | 8.76 | 8.74 |
| 4 | 7.71 | 6.94 | 6.59 | 6.39 | 6.26 | 6.16 | 6.09 | 6.04 | 6.00 | 5.96 | 5.93 | 5.91 |
| 5 | 6.61 | 5.79 | 5.41 | 5.19 | 5.05 | 4.95 | 4.88 | 4.82 | 4.77 | 4.74 | 4.71 | 4.68 |
| 6 | 5.99 | 5.14 | 4.76 | 4.53 | 4.39 | 4.28 | 4.21 | 4.15 | 4.10 | 4.06 | 4.03 | 4.00 |
| 7 | 5.59 | 4.74 | 4.35 | 4.12 | 3.97 | 3.87 | 3.77 | 3.73 | 3.68 | 3.64 | 3.60 | 3.57 |
| 8 | 5.32 | 4.46 | 4.07 | 3.84 | 3.69 | 3.58 | 3.50 | 3.44 | 3.39 | 3.35 | 3.31 | 3.28 |
| 9 | 5.12 | 4.26 | 3.86 | 3.63 | 3.48 | 3.37 | 3.29 | 3.23 | 3.18 | 3.14 | 3.10 | 3.07 |
| 10 | 4.96 | 4.10 | 3.71 | 3.48 | 3.33 | 3.22 | 3.14 | 3.07 | 3.02 | 2.98 | 2.94 | 2.91 |
| 11 | 4.84 | 3.98 | 3.59 | 3.36 | 3.20 | 3.09 | 3.01 | 2.95 | 2.90 | 2.85 | 2.82 | 2.79 |
| 12 | 4.75 | 3.89 | 3.49 | 3.26 | 3.11 | 3.00 | 2.91 | 2.85 | 2.80 | 2.75 | 2.72 | 2.69 |
| 15 | 4.54 | 3.68 | 3.29 | 3.06 | 2.90 | 2.79 | 2.71 | 2.64 | 2.59 | 2.54 | 2.51 | 2.48 |
| 20 | 4.35 | 3.49 | 3.10 | 2.87 | 2.71 | 2.60 | 2.51 | 2.45 | 2.39 | 2.35 | 2.31 | 2.28 |
| 25 | 4.24 | 3.39 | 2.99 | 2.76 | 2.60 | 2.49 | 2.40 | 2.34 | 2.28 | 2.24 | 2.21 | 2.16 |
| 30 | 4.17 | 3.32 | 2.92 | 2.69 | 2.53 | 2.42 | 2.33 | 2.27 | 2.21 | 2.16 | 2.13 | 2.09 |
| 40 | 4.08 | 3.23 | 2.84 | 2.61 | 2.45 | 2.34 | 2.25 | 2.18 | 2.12 | 2.08 | 2.04 | 2.04 |
| 60 | 4.00 | 3.15 | 2.76 | 2.53 | 2.37 | 2.25 | 2.17 | 2.10 | 2.04 | 1.99 | 1.95 | 1.92 |
| 120 | 3.92 | 3.07 | 2.68 | 2.45 | 2.29 | 2.17 | 2.09 | 2.02 | 1.96 | 1.91 | 1.87 | 1.83 |

**Table A.6** Part B: $\alpha = 0.01$

| | $df_1$ | | | | | | | | | | | |
|---|---|---|---|---|---|---|---|---|---|---|---|---|
| $df_2$ | 1 | 2 | 3 | 4 | 5 | 6 | 7 | 8 | 9 | 10 | 11 | 12 |
| 2 | 98.5 | 99.0 | 99.2 | 99.2 | 99.3 | 99.3 | 99.4 | 99.4 | 99.4 | 99.4 | 99.4 | 99.4 |
| 3 | 34.1 | 30.8 | 29.5 | 28.7 | 28.2 | 27.9 | 27.7 | 27.5 | 27.3 | 27.2 | 27.1 | 27.1 |
| 4 | 21.2 | 18.0 | 16.7 | 16.0 | 15.5 | 15.2 | 15.0 | 14.8 | 14.7 | 14.5 | 14.4 | 14.4 |
| 5 | 16.3 | 13.3 | 12.1 | 11.4 | 11.0 | 10.7 | 10.5 | 10.3 | 10.2 | 10.1 | 9.99 | 9.89 |
| 6 | 13.7 | 10.9 | 9.78 | 9.15 | 8.75 | 8.47 | 8.26 | 8.10 | 7.98 | 7.87 | 7.79 | 7.72 |
| 7 | 12.2 | 9.55 | 8.45 | 7.85 | 7.46 | 7.19 | 6.99 | 6.84 | 6.72 | 6.62 | 6.54 | 6.47 |
| 8 | 11.3 | 8.65 | 7.59 | 7.01 | 6.63 | 6.37 | 6.18 | 6.03 | 5.91 | 5.81 | 5.73 | 5.67 |
| 9 | 10.6 | 8.02 | 6.99 | 6.42 | 6.06 | 5.80 | 5.61 | 5.47 | 5.35 | 5.26 | 5.18 | 5.11 |
| 10 | 10.0 | 7.56 | 6.55 | 5.99 | 5.64 | 5.39 | 5.20 | 5.06 | 4.94 | 4.85 | 4.77 | 4.71 |
| 11 | 9.65 | 7.21 | 6.22 | 5.67 | 5.32 | 5.07 | 4.89 | 4.74 | 4.63 | 4.54 | 4.46 | 4.40 |
| 12 | 9.33 | 6.93 | 5.95 | 5.41 | 5.06 | 4.82 | 4.64 | 4.50 | 4.39 | 4.30 | 4.22 | 4.16 |
| 15 | 8.68 | 6.36 | 5.42 | 4.89 | 4.56 | 4.32 | 4.14 | 4.00 | 3.89 | 3.80 | 3.73 | 3.67 |
| 20 | 8.10 | 5.85 | 4.94 | 4.43 | 4.10 | 3.87 | 3.70 | 3.56 | 3.46 | 3.37 | 3.29 | 3.23 |
| 25 | 7.77 | 5.57 | 4.68 | 4.18 | 3.86 | 3.63 | 3.46 | 3.32 | 3.22 | 3.13 | 3.06 | 2.99 |
| 30 | 7.56 | 5.39 | 4.51 | 4.02 | 3.70 | 3.47 | 3.30 | 3.17 | 3.07 | 2.98 | 2.90 | 2.84 |
| 40 | 7.31 | 5.18 | 4.31 | 3.83 | 3.51 | 3.29 | 3.12 | 2.99 | 2.89 | 2.80 | 2.73 | 2.66 |
| 60 | 7.08 | 4.98 | 4.13 | 3.65 | 3.34 | 3.12 | 2.95 | 2.82 | 2.72 | 2.63 | 2.56 | 2.50 |
| 120 | 6.85 | 4.79 | 3.95 | 3.48 | 3.17 | 2.96 | 2.79 | 2.66 | 2.56 | 2.47 | 2.40 | 2.34 |

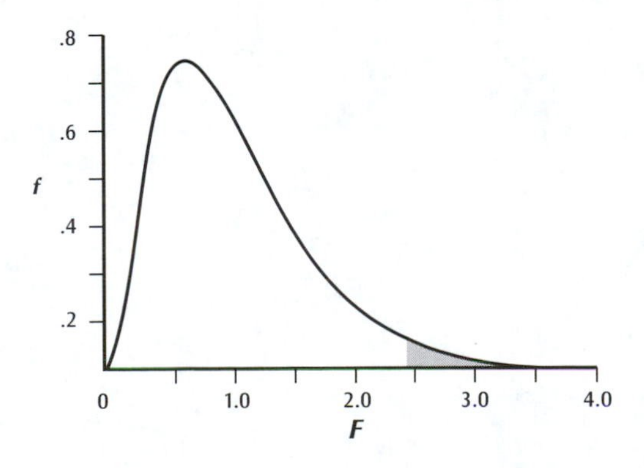

**Table A.7** Critical values of $q$ (Studentized $t$) for the Tukey test ($\alpha = 0.05$)

| Error df | Number of Groups (Treatments) | | | | | | | | |
|---|---|---|---|---|---|---|---|---|---|
| | 2 | 3 | 4 | 5 | 6 | 7 | 8 | 9 | 10 |
| 1 | 17.97 | 26.98 | 32.82 | 37.08 | 40.41 | 43.12 | 45.40 | 47.36 | 49.07 |
| 2 | 6.08 | 8.33 | 9.80 | 10.88 | 11.74 | 12.44 | 13.03 | 13.54 | 13.99 |
| 3 | 4.50 | 5.91 | 6.82 | 7.50 | 8.04 | 8.48 | 8.85 | 9.18 | 9.46 |
| 4 | 3.93 | 5.04 | 5.76 | 6.29 | 6.71 | 7.05 | 7.35 | 7.60 | 7.83 |
| 5 | 3.64 | 4.60 | 5.22 | 5.67 | 6.03 | 6.33 | 6.58 | 6.80 | 6.99 |
| 6 | 3.46 | 4.34 | 4.90 | 5.30 | 5.63 | 5.90 | 6.12 | 6.32 | 6.49 |
| 7 | 3.34 | 4.16 | 4.68 | 5.06 | 5.36 | 5.61 | 5.82 | 6.00 | 6.16 |
| 8 | 3.26 | 4.04 | 4.53 | 4.89 | 5.17 | 5.40 | 5.60 | 5.77 | 5.92 |
| 9 | 3.20 | 3.95 | 4.41 | 4.76 | 5.02 | 5.24 | 5.43 | 5.59 | 5.74 |
| 10 | 3.15 | 3.88 | 4.33 | 4.65 | 4.91 | 5.12 | 5.30 | 5.46 | 5.60 |
| 11 | 3.11 | 3.82 | 4.26 | 4.57 | 4.82 | 5.03 | 5.20 | 5.35 | 5.49 |
| 12 | 3.08 | 3.77 | 4.20 | 4.51 | 4.75 | 4.95 | 5.12 | 5.27 | 5.39 |
| 13 | 3.06 | 3.73 | 4.15 | 4.45 | 4.69 | 4.88 | 5.05 | 5.19 | 5.32 |
| 14 | 3.03 | 3.70 | 4.11 | 4.41 | 4.64 | 4.83 | 4.99 | 5.13 | 5.25 |
| 15 | 3.01 | 3.67 | 4.08 | 4.37 | 4.59 | 4.78 | 4.94 | 5.08 | 5.20 |
| 16 | 3.00 | 3.65 | 4.05 | 4.33 | 4.56 | 4.74 | 4.90 | 5.03 | 5.15 |
| 17 | 2.98 | 3.63 | 4.02 | 4.30 | 4.52 | 4.70 | 4.86 | 4.99 | 5.11 |
| 18 | 2.97 | 3.61 | 4.00 | 4.28 | 4.49 | 4.67 | 4.82 | 4.96 | 5.07 |
| 19 | 2.96 | 3.59 | 3.98 | 4.25 | 4.47 | 4.65 | 4.79 | 4.92 | 5.04 |
| 20 | 2.95 | 3.58 | 3.96 | 4.23 | 4.45 | 4.62 | 4.77 | 4.90 | 5.01 |
| 24 | 2.92 | 3.53 | 3.90 | 4.17 | 4.37 | 4.54 | 4.68 | 4.81 | 4.92 |
| 30 | 2.89 | 3.49 | 3.85 | 4.10 | 4.30 | 4.46 | 4.60 | 4.72 | 4.82 |
| 40 | 2.86 | 3.44 | 3.79 | 4.04 | 4.23 | 4.39 | 4.52 | 4.63 | 4.73 |
| 60 | 2.83 | 3.40 | 3.74 | 3.98 | 4.16 | 4.31 | 4.44 | 4.55 | 4.65 |
| 120 | 2.80 | 3.36 | 3.68 | 3.92 | 4.10 | 4.24 | 4.36 | 4.47 | 4.56 |
| ∞ | 2.77 | 3.31 | 3.63 | 3.86 | 4.03 | 4.17 | 4.29 | 4.39 | 4.47 |

**Table A.8** Critical values of the Pearson correlation coefficient ($r$) ($\alpha$ = 0.05)*

| df | r |
|---|---|
| 1 | 0.997 |
| 2 | 0.950 |
| 3 | 0.878 |
| 4 | 0.811 |
| 5 | 0.754 |
| 6 | 0.707 |
| 7 | 0.666 |
| 8 | 0.632 |
| 9 | 0.602 |
| 10 | 0.576 |
| 11 | 0.553 |
| 12 | 0.532 |
| 13 | 0.514 |
| 14 | 0.497 |
| 15 | 0.482 |
| 16 | 0.468 |
| 17 | 0.456 |
| 18 | 0.444 |
| 19 | 0.433 |
| 20 | 0.423 |
| 21 | 0.413 |
| 22 | 0.404 |
| 23 | 0.396 |
| 24 | 0.388 |
| 25 | 0.381 |
| 26 | 0.374 |
| 27 | 0.367 |
| 28 | 0.361 |
| 29 | 0.355 |
| 30 | 0.349 |
| 35 | 0.325 |
| 40 | 0.304 |
| 45 | 0.288 |
| 50 | 0.273 |
| 60 | 0.250 |
| 70 | 0.232 |
| 80 | 0.217 |
| 90 | 0.205 |
| 100 | 0.195 |
| 120 | 0.174 |

*For sample sizes not tabulated, either extrapolate or test for significance using the $t$-test.

**Table A.9** Critical values of the Spearman rank correlation coefficient ($\alpha$ = 0.05)*

| $n$ | $r_s$ |
|---|---|
| 4 | 1.000 |
| 5 | 0.900 |
| 6 | 0.829 |
| 7 | 0.714 |
| 8 | 0.643 |
| 9 | 0.600 |
| 10 | 0.564 |
| 12 | 0.506 |
| 14 | 0.456 |
| 16 | 0.425 |
| 18 | 0.399 |
| 20 | 0.377 |

*For sample sizes not tabulated, test for significance using the $t$-test.

**Table A.10** Table of random numbers

| 874335 | 218040 | 632420 | 240295 | 301131 | 152740 | 433058 | 274170 |
|---|---|---|---|---|---|---|---|
| 142131 | 051859 | 719342 | 714391 | 174251 | 147150 | 108520 | 771712 |
| 577728 | 460401 | 847722 | 767239 | 201744 | 006565 | 204589 | 960553 |
| 080052 | 246887 | 107893 | 627841 | 196599 | 792021 | 038162 | 390011 |
| 501153 | 355165 | 168311 | 790826 | 174928 | 955178 | 754258 | 125025 |
| 146207 | 369709 | 775557 | 516449 | 855970 | 838321 | 826020 | 246163 |
| 273515 | 015616 | 254341 | 330587 | 162088 | 174360 | 554720 | 349616 |
| 594504 | 658609 | 007492 | 524747 | 718771 | 586831 | 569750 | 047201 |
| 773722 | 805035 | 015969 | 656055 | 354632 | 089893 | 328631 | 466358 |
| 928848 | 601866 | 338853 | 047266 | 601409 | 588331 | 617007 | 750155 |
| 130680 | 336701 | 613351 | 286758 | 193966 | 377556 | 048648 | 557283 |
| 903145 | 937763 | 554796 | 728537 | 570290 | 643603 | 565449 | 562057 |
| 723294 | 473898 | 456644 | 992231 | 371495 | 963132 | 937428 | 954420 |
| 521302 | 654580 | 690478 | 463092 | 941820 | 803428 | 262731 | 939938 |
| 180471 | 329905 | 206005 | 792002 | 828627 | 022402 | 467626 | 239803 |
| 037226 | 990598 | 031055 | 395463 | 282404 | 368588 | 806509 | 590830 |
| 381118 | 268005 | 771588 | 955604 | 756766 | 981147 | 361899 | 245461 |
| 954822 | 434100 | 111684 | 920179 | 408451 | 889864 | 544440 | 471762 |
| 454139 | 901479 | 313550 | 002567 | 597321 | 515148 | 592903 | 053426 |
| 027996 | 723365 | 717520 | 681773 | 386364 | 168036 | 074181 | 789768 |
| 778443 | 093607 | 242049 | 702424 | 041696 | 550187 | 383294 | 995730 |
| 260656 | 846676 | 883719 | 574775 | 532552 | 253887 | 243386 | 001878 |
| 982935 | 957671 | 217239 | 074705 | 031298 | 262045 | 205728 | 654403 |
| 906706 | 042314 | 895439 | 743718 | 413420 | 448197 | 149714 | 815122 |
| 946521 | 856953 | 149277 | 388942 | 757533 | 076503 | 782862 | 861477 |
| 470054 | 798560 | 287835 | 583131 | 845375 | 301748 | 140819 | 186534 |
| 798107 | 404733 | 198320 | 164665 | 661808 | 669342 | 087352 | 698984 |
| 704605 | 853694 | 846064 | 737547 | 894822 | 615321 | 814358 | 323143 |
| 916600 | 464292 | 774523 | 171407 | 435529 | 966344 | 341855 | 498953 |
| 614267 | 196000 | 605281 | 101497 | 878168 | 439697 | 017987 | 681981 |
| 930906 | 148913 | 538043 | 428698 | 020102 | 143290 | 019025 | 843417 |
| 452944 | 063756 | 850643 | 819512 | 361819 | 075658 | 849363 | 970079 |
| 719931 | 821876 | 399037 | 206069 | 606933 | 625961 | 841521 | 564408 |
| 724544 | 945246 | 117307 | 286123 | 162181 | 073984 | 656142 | 144469 |
| 412582 | 096463 | 517660 | 023052 | 637428 | 090138 | 781997 | 743955 |
| 182972 | 578750 | 190428 | 145861 | 345662 | 235457 | 035980 | 412182 |
| 387765 | 835955 | 304068 | 649179 | 802995 | 461602 | 063111 | 714091 |
| 832135 | 952549 | 105163 | 293258 | 228666 | 610859 | 836534 | 230248 |
| 274385 | 153632 | 418418 | 103979 | 045038 | 916136 | 157518 | 056846 |
| 925940 | 304925 | 146667 | 872845 | 377600 | 500970 | 155459 | 305700 |

# Appendix B

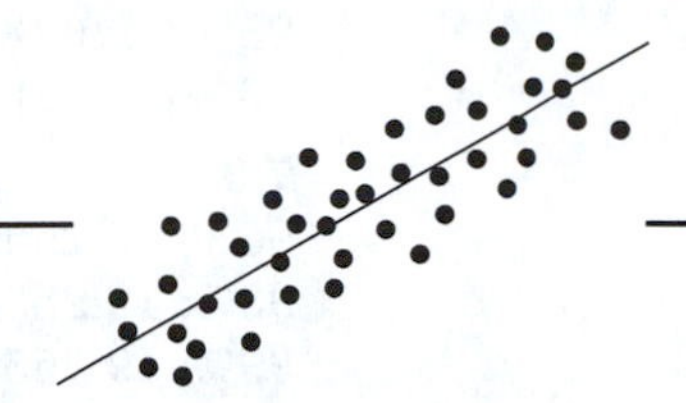

# Answers to Odd-Numbered and Selected Exercises

Answers to selected exercises starting with chapter 2 are shown below. Rounding errors may cause slight discrepancies between these answers and your answers.

## CHAPTER 2

**2.1** a) ratio b) continuous

**2.3** a) ratio b) discrete

**2.5–2.6**

| Male Number | 1 | 2 | 3 | 4 | 5 | 6 |
|---|---|---|---|---|---|---|
| Height | 181 | 202 | 190 | 185 | 190 | 200 |
| Rank | 1 | 6 | 3.5 | 2 | 3.5 | 5 |
| Category | S | T | S | S | S | T |

**2.7–2.8**

| Pond Number | 1 | 2 | 3 | 4 | 5 | 6 | 7 | 8 | 9 | 10 | 11 |
|---|---|---|---|---|---|---|---|---|---|---|---|
| No. Bullfrogs | 34 | 65 | 23 | 34 | 18 | 20 | 15 | 70 | 15 | 18 | 34 |
| Rank | 8.0 | 10 | 6 | 8.0 | 3.5 | 5 | 1.5 | 11 | 1.5 | 3.5 | 8.0 |
| Category | D | D | S | D | S | S | S | D | S | S | D |

**2.9**

a) continuous measurement, ratio, continuous, derived

b) discrete measurement, ratio, discrete, simple

c) attribute, nominal

d) continuous measurement, ratio, continuous, transformed

e) continuous measurement, ratio, continuous, simple

## CHAPTER 4

**4.1** Mean 6.0; Median 6.5; Range 1, 11; IQ range 3.0, 9.0; Variance 8.9655; Standard Deviation 2.9942; CV 49.9%

**4.3** Mean 55.6976; Median 56.5; Range 47.0, 62.4; IQ range 52.175, 59.225; Variance 18.4310; Standard Deviation 4.2931; CV 7.7%

**4.5** Mean 274.2100

**4.9** Median = 2

## CHAPTER 5

**5.1** 0.25

**5.3** 0.50

**5.5** 0.34375

**5.7** 0.33021

**5.9** 0.3125

**5.11** 0.015625

**5.13** 0.177979

**5.15** 0.822021

**5.17**
P(0 males) = 0.004577, P(1) = 0.035178, . . . . . ., P(8) = 0.003323

**5.19** $3.323 \approx 3$

**5.21**
mean = 2.99 $\Sigma f = 100$

| $x$ | $p(x)$ | $f(x)$ |
|---|---|---|
| 0 | 0.0503 | 5.03 |
| 1 | 0.1504 | 15.04 |
| 2 | 0.2248 | 22.48 |
| 3 | 0.2240 | 22.40 |
| 4 | 0.1675 | 16.75 |
| 5 | 0.1001 | 10.01 |
| 6 | 0.0499 | 4.99 |
| 7 | 0.0213 | 2.13 |
| 8 | 0.0080 | 0.80 |

**5.23** a) 0.007967 b) 0.179336

**5.25** $\mu$ = 0.6 phage/cell; $p\ (\geq 3) = 0.0231$

## CHAPTER 6

**6.1** 0.0021

**6.3** 0.6331

**6.5** 0.1814

**6.7** 0.2386

**6.9** 0.0968

**6.11** 0.3707

**6.13** 0.8060

## CHAPTER 7

**7.1–7.6** [Answers depend on random samples collected]

**7.7**
90% CI: 6.00 ± 0.93 = 5.07, 6.93
95% CI: 6.00 ± 1.12 = 4.88, 7.12
99% CI: 6.00 ± 1.51 = 4.49, 7.51

**7.9**
90%; 19.2556 ± 1.5895
95%; 19.2556 ± 1.9707
99%; 19.2556 ± 2.8672

**7.11**
90%; 45.1818 ± 5.4473
95%; 45.1818 ± 6.6967
99%; 45.1818 ± 9.5252

**7.13**
90%; 4.9760 ± 0.2877
95%; 4.9760 ± 0.3471
99%; 4.9760 ± 0.4703

## CHAPTER 8

**8.1**
$H_0: \mu \leq 10$ ppb vs. $H_a: \mu > 10$ ppb
$t = 2.50$, one-tailed $p$ value: $0.005 < p < 0.01$
reject $H_0$, exceeds 10 ppb

**8.3**
$H_0: \mu = 100$ vs. $H_a: \mu \neq 100$
$|t| = 5.367, p < 0.0001$, reject $H_0$; mean weight differs from 100; so this hen should be removed from the breeding program

**8.5**
$H_0: \mu \geq 425$ vs. $H_a: \mu < 425$
$t = 0.78$, $p > 0.20$, fail to reject, not less than 425 (ponds not lower than lakes)

## CHAPTER 9

**9.1**
$H_0$: $\mu$(men) = $\mu$(women) vs. $H_a$: $\mu$(men) ≠ $\mu$(women)
$|t| = 3.83$, $p$ value $0.0001 < p < 0.001$, reject $H_0$, genders are different
95% CI for $\mu_w - \mu_m$ = 8.481 ± 4.437 = 4.044, 12.918
We are 95% sure that the resting pulse rate is between 4.0 and 12.9 beats per minute greater in women than in men.

**9.3**
$H_0$: $\mu$(largemouth) = $\mu$(smallmouth) vs. $H_a$: $\mu$(largemouth) ≠ $\mu$(smallmouth)
$|t| = 11.33$

**9.5**
$H_0$: $\mu$(down) ≤ $\mu$(up) vs. $H_a$: $\mu$(down) > $\mu$(up)
$t = 5.14$

**9.7**
$H_0$: $\mu$(glucose) = $\mu$(sucrose) vs. $H_a$: $\mu$(glucose) ≠ $\mu$(sucrose)
$t = 0.731$

**9.9**
$H_0$: $\mu$(with) ≤ $\mu$(without) vs. $H_a$: $\mu$(with) > $\mu$(without)
$t = 6.1612$

**9.11**
$H_0$: $\mu$(coniferous) = $\mu$(deciduous) vs. $H_a$: $\mu$(coniferous) ≠ $\mu$(deciduous)
$t = 4.1210$

**9.13**
$H_0$: $\mu$(smokers) ≤ $\mu$(non) vs. $H_a$: $\mu$(smokers) > $\mu$(non)
$t = -0.67$

**9.15**
$H_0$: $\theta$(hairy) ≤ $\theta$(non-hairy) vs. $H_a$: $\theta$(hairy) > $\theta$(non-hairy)
$U = 91.5$, $U_{crit} = 73$, $p << 0.05$, reject $H_0$, hairy appear scarier!

**9.17** $U = 123.5$, $p < 0.05$

**9.19** $U = 57$, $U_{crit} = 71$, $p > 0.05$

**9.21**
$H_0$: $\mu$(difference) = 0 vs. $H_a$: $\mu$(difference) ≠ 0
$t = 4.27$

**9.23**
$H_0$: $\mu$(wall – center) ≤ 0 vs. $H_a$: $\mu$(wall – center) > 0
$t = 2.3904$

**9.25** $T = 0$

**9.27** $p = 0.0625$

**9.29** $U = 90.5$, $p > 0.05$

**9.31**
$H_0$: $\mu$(difference) = 0 vs. $H_a$: $\mu$(difference) ≠ 0
$t = 2.94$, $0.01 < p < 0.02$

**9.33**
$H_0$: $\mu$(infected) = $\mu$(healthy)
$H_1$: $\mu$(infected) ≠ $\mu$(healthy)
$t = 3.71$

## CHAPTER 10

**10.1**
$H_0$: all $\mu$'s equal vs. $H_a$: one or more $\mu$'s different
$F_{2,12,(.05)} = 3.89$, $F_{stat} = 30.87$, $p << 0.05$, reject $H_0$

| Source | Sum of Squares | df | MS | *F* |
|---|---|---|---|---|
| Between-Groups | 6.2713 | 2 | 3.1356 | 30.87 |
| Error | 1.2191 | 12 | 0.1016 | |
| Total | 7.4904 | 14 | | |

**10.3**

| Source | Sum of Squares | df | MS | *F* |
|---|---|---|---|---|
| Between-Groups | 632.95 | 3 | 210.98 | 70.33 |
| Error | 48.00 | 16 | 3.00 | |
| Total | 680.95 | 19 | | |

**10.5**

| Source | Sum of Squares | df | MS | *F* |
|---|---|---|---|---|
| Between-Groups | 1740.4 | 2 | 870.2 | 79.11 |
| Error | 132.0 | 12 | 11.0 | |
| Total | 1872.4 | 14 | | |

**10.7**

| Source | Sum of Squares | df | MS | *F* |
|---|---|---|---|---|
| Between-Groups | 3785.65 | 4 | 946.41 | 72.78 |
| Error | 455.13 | 35 | 13.00 | |
| Total | 4240.78 | 39 | | |

**10.9**

| Source | Sum of Squares | df | MS | *F* |
|---|---|---|---|---|
| Between-Groups | 145.6389 | 3 | 48.5463 | 19.64 |
| Error | 79.1111 | 32 | 2.4722 | |
| Total | 224.7500 | 35 | | |

**10.11–10.18**
Using MINITAB, homogeneity of variance test suggests all except 10.17 and 10.18 conform to assumptions of equal variance.

**10.19**
$H_0$: all $\theta$'s equal vs. $H_a$: one or more $\theta$'s different
$H = 9.39$, $\chi^2_{2,.05} = 5.99$, $p < 0.01$, reject $H_0$.

**10.21** $H = 12.42$

**10.23** $H = 10.38$

# CHAPTER 11

**11.1**

$H_0$: no difference among mercury treatments vs. $H_a$: different

$F_{2,10,(.05)} = 4.10$

$H_0$: no variation among litters vs. $H_a$: variation among litters

$F_{5,10,(0.05)} = 3.33$

| Source | Sum of Squares | df | MS | $F$ |
|---|---|---|---|---|
| Mercury | 3169.44 | 2 | 1584.72 | 15.63 |
| Blocks (litters) | 15573.61 | 5 | 3114.72 | 30.72 |
| Error | 1013.89 | 10 | 101.39 | |
| Total | 19756.94 | 17 | | |

Conclusion: both mercury and litter affect activity levels

**11.3**

| Source | Sum of Squares | df | MS | $F$ |
|---|---|---|---|---|
| Treatment | 1044.40 | 2 | 522.20 | 14.16 |
| Blocks (plant) | 39876.27 | 4 | 9969.07 | 270.41 |
| Error | 294.13 | 8 | 36.87 | |
| Total | 41215.60 | 14 | | |

**11.5**

| Source | Sum of Squares | df | MS |
|---|---|---|---|
| Between-Groups | 587.56 | 2 | 293.78 |
| Blocks | 389.56 | 2 | 194.78 |
| Error | 40.44 | 4 | 10.11 |
| Total | 1017.56 | 8 | |

**11.7**

| Source | Sum of Squares | df | MS |
|---|---|---|---|
| Between-Groups | 178188.50 | 2 | 89094.25 |
| Blocks | 84247.00 | 3 | 28082.33 |
| Error | 30321.50 | 6 | 5053.58 |
| Total | 292757.00 | 11 | |

**11.9**

$H_0$: no difference between sex/reproductive groups vs. $H_a$: different

$F_{2,45,(.05)} = 3.21$

$H_0$: no effect of crowding vs. $H_a$: crowding effect

$F_{2,45,(.05)} = 3.21$

$H_0$: no interaction between sex and crowding vs. $H_a$: interaction effect

$F_{4,45,(.05)} = 2.59$

| Source | Sum of Squares | df | MS | $F$ |
|---|---|---|---|---|
| Sex | 473126 | 2 | 236563 | 1440.73 |
| Stress | 16864 | 2 | 8432 | 51.35 |
| Interaction | 4401 | 4 | 1100 | 6.70 |
| Error | 7389 | 45 | 164 | |
| Total | 501780 | 53 | | |

Decisions: reject all three null hypotheses

**11.11**

| Source | Sum of Squares | df | MS | $F$ |
|---|---|---|---|---|
| Age | 6661.67 | 2 | 3330.83 | 46.9 |
| Sex | 616.53 | 1 | 616.53 | 8.68 |
| Interaction | 622.07 | 2 | 311.03 | 4.38 |
| Error | 1704.40 | 24 | 71.02 | |
| Total | 9604.67 | 29 | | |

**11.13**

$\chi^2 = 22.1$

$\chi^2_{4,(.05)} = 9.49$, $0.0001 < p < 0.001$

**11.15**

| Source | Sum of Squares | df | MS | $F$ |
|---|---|---|---|---|
| Caffeine | 483.44 | 2 | 241.72 | 110.43 |
| Blocks (rats) | 1369.11 | 5 | 273.82 | 125.03 |
| Error | 21.89 | 10 | 2.19 | |
| Total | 1874.44 | 17 | | |

**11.17**

| Source | Sum of Squares | df | MS | $F$ |
|---|---|---|---|---|
| Strain | 19683.63 | 2 | 9841.81 | 33.3 |
| Treatment | 58569.85 | 2 | 29284.93 | 99.07 |
| Interaction | 407.23 | 4 | 101.81 | 0.34 |
| Error | 5320.67 | 18 | 295.59 | |
| Total | 83981.41 | 26 | | |

**11.19**

| Source | Sum of Squares | df | MS | $F$ |
|---|---|---|---|---|
| Between-Groups | 95033.83 | 3 | 31677.94 | 27.71 |
| Blocks | 2460.83 | 5 | 492.17 | |
| Error | 17149.17 | 15 | 1143.28 | |
| Total | 114643.83 | 23 | | |

## CHAPTER 12

**12.1**
The plot indicates a tight linear relationship.
$\hat{y} = -1.90 + 10.7x$; $r^2 = 0.999$

| Source | SS | df | MS | F |
|---|---|---|---|---|
| Regression | 33093 | 1 | 33093 | 4750.44 |
| Error | 28 | 4 | 7 | |
| Total | 33121 | 5 | | |

$H_0$: $\beta = 0$ vs. $H_a$: $\beta \neq 0$; $F_{1,4,(.05)} = 7.71$; reject $H_0$

| $x$ | $\hat{y}$ | ± | 95% CI for $\hat{y}$ | |
|---|---|---|---|---|
| 0 | −1.9 | 4.058182 | −5.95818 | 2.158182 |
| 1 | 8.7947 | 3.780703 | 5.013997 | 12.5754 |
| 2 | 19.4894 | 3.534326 | 15.95507 | 23.02373 |
| 5 | 51.5735 | 3.053193 | 48.52031 | 54.62669 |
| 10 | 105.047 | 3.39074 | 101.6563 | 108.4377 |
| 20 | 211.994 | 6.619165 | 205.3748 | 218.6132 |

**12.3**
$\hat{y} = 100 + 3.90x$; $r^2 = 0.909$

| Source | SS | df | MS | F |
|---|---|---|---|---|
| Regression | 912.60 | 1 | 912.60 | 70.23 |
| Error | 90.96 | 7 | 12.99 | |
| Total | 1003.56 | 8 | | |

**12.5**
Let $y = 1/\text{rate}$, $x = 1/\text{concentration}$
$\hat{y} = 0.00258 + 0.00137x$; $r^2 = 0.996$

| Source | SS | df | MS | F |
|---|---|---|---|---|
| Regression | 0.000017103 | 1 | 0.000017103 | 836.85 |
| Error | 0.000000061 | 3 | 0.000000020 | |
| Total | 0.000017164 | 4 | | |

**12.7**
$\hat{y} = 15.1 + 1.46x$; $r^2 = 0.932$

| Source | SS | df | MS | F |
|---|---|---|---|---|
| Regression | 235.65 | 1 | 235.65 | 124.16 |
| Error | 17.08 | 9 | 1.90 | |
| Total | 252.73 | 10 | | |

**12.9**
$\hat{y} = 6.52 + 0.811x$; $r^2 = 0.909$

| Source | SS | df | MS | F |
|---|---|---|---|---|
| Regression | 2137.0 | 1 | 2137.0 | 240.1 |
| Error | 213.6 | 24 | 8.9 | |
| Total | 2350.6 | 25 | | |

## CHAPTER 13

**13.1** $r = 0.827$

**13.3**
age vs. systolic: r = 0.590
age vs. diastolic: r = 0.590
systolic vs. diastolic: r = 0.747

**13.5** $r = 0.994$

**13.7**
weight vs. width: $r = 0.628$
weight vs. length: $r = 0.618$
width vs. length: $r = 0.221$

**13.9**
$\Sigma d^2 = 44$, $r_s = 0.473$, $r_{8,(.05)} = 0.643$, correlation is not significant

## CHAPTER 14

**14.1** $\chi^2 = 25.0$, $p < 0.0001$

**14.3** $\chi^2 = 0.333$

**14.5** $\chi^2 = 1.33$

**14.7**
$\chi^2 = 26.78$ (The first 4 categories should be combined, giving 2 degrees of freedom.)

**14.9**
$\chi^2 = 0.2023$ (The last 3 categories should be combined, giving 6 degrees of freedom.)

**14.11** $\chi^2 = 12.04$

Exercises 14.12 through 14.13 are intended to be done by computer.

**14.15**
One-tailed $p = 0.06555$ $(0.04895 + 0.01398 + 0.00233 + 0.00029)$

**14.17**
Pain relief: $\chi^2 = 17.282$, $\chi^2_{2,(.05)} = 5.99$, $0.0001 < p < 0.001$
Stiffness relief: $\chi^2 = 11.618$

**14.19** $\chi^2 = 3.887$

**14.21** $\chi^2 = 32.06$

**14.23** $\chi^2 = 18.529$, $\chi^2_{2,(.05)} = 5.99$, $p < 0.0001$

# Appendix C

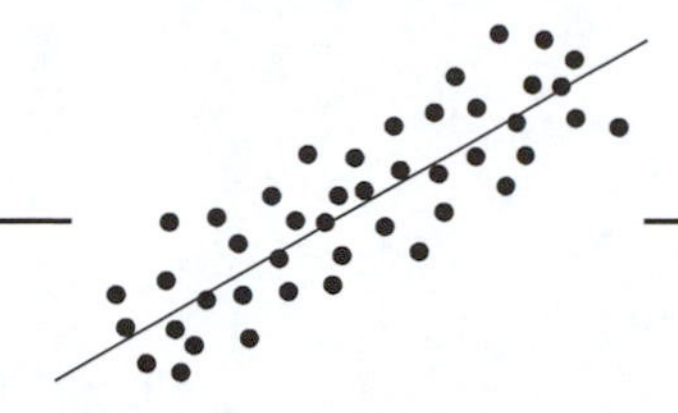

# References on Statistics, Experimental Design, and Applications


Ambrose, H. W., and K. P. Ambrose. 1981. *A Handbook of Biological Investigation.* 3rd ed. Hunter Textbooks.

Box, G. E. P., W. G. Hunter, and J. S. Hunter. 1978. *Statistics for Experimenters: An Introduction to Design, Data Analysis, and Model Building.* John Wiley & Sons.

Forthofer, R. N., and E. S. Lee. 1995. *Introduction to Biostatistics: A Guide to Design, Analysis, and Discovery.* Academic Press.

Gotelli, N. J., and A. M. Ellison. 2004. *A Primer of Ecological Statistics.* Sinauer.

Green, R. H. 1979. *Sampling Design and Statistical Methods for Environmental Biologists.* John Wiley & Sons.

Hollander, M., and D. A. Wolfe. 1973. *Nonparametric Statistical Methods.* John Wiley & Sons.

Lilienfeld, A. M., and D. E. Lilienfeld. 1980. *Foundations of Epidemiology.* 2nd ed. Oxford University Press.

McGarigal, K., S. Cushman, and S. Stafford. 2000. *Multivariate Statistics for Wildlife and Ecology Research.* Springer-Verlag.

Mead, R. 1988. *The Design of Experiments: Statistical Principles for Practical Application.* Cambridge University Press.

Neter, J., W. Wasserman, and M. H. Kutner. 1989. *Applied Linear Regression Models.* 2nd ed. Irwin.

Quinn, G. P., and M. J. Keough. 2002. *Experimental Design and Data Analysis for Biologists.* Cambridge University Press.

Scheiner, S. M., and J. Gurevitch (eds.). 1993. *Design and Analysis of Ecological Experiments.* Chapman and Hall.

Siegel, S. 1956. *Nonparametric Statistics for the Behavioral Sciences.* McGraw-Hill.

Siegel, S., and N. J. Castellan. 1988. *Nonparametric Statistics for the Behavioral Sciences.* 2nd ed. McGraw-Hill.

Sokal, R. R., and J. J. Rohlf. 1995. *Biometry.* 3rd ed. Freeman.

Snedecor, G. W., and W. G. Cochran. 1980. *Statistical Methods.* 7th ed. Iowa State University Press

Stevens, J. 1996. *Applied Multivariate Statistics for the Social Sciences.* 3rd ed. Lawrence Erlbaum Associates.

Wonnacott, T. H., and R. J. Wonnacott. 1977. *Introductory Statistics for Business and Economics.* 2nd ed. John Wiley & Sons.

Zar, J. H. 1999. *Biostatistical Analysis.* 4th ed. Prentice Hall.

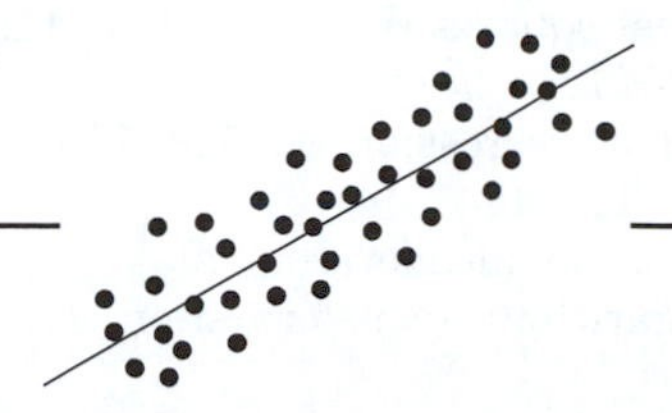

# Index